WORK TRANSFORMED

Harley Shaiken

Automation and Labor in the Computer Age

HOLT, RINEHART AND WINSTON
New York

First published in February 1985 by Holt, Rinehart and Winston,
383 Madison Avenue, New York, New York 10017.
Published simultaneously in Canada by Holt, Rinehart and
Winston of Canada, Limited.

Library of Congress Cataloging in Publication Data
Shaiken, Harley.
Work transformed.
Includes bibliographical references and index.
1. Labor supply—Effect of automation on. I. Title.
HD6331.S475 1985 331.25 84-4675
ISBN 0-03-042681-2

First Edition

Design by Dalia Bergsagel
Printed in the United States of America
1 3 5 7 9 10 8 6 4 2

ISBN 0-03-042681-2

To Bickie and Mariela

CONTENTS

Acknowledgments ix

Preface xi

1. Introduction 1

2. The Machine Shop 16
   Scientific Management
   The Moral Code

3. Criteria of Design 45

4. Numerical Control: A Case Study 66
   Technical and Economic Advantages
   Numerical Control on the Shop Floor
   Programming and Computer Numerical Control
   The Impact on the Machinist

5. The Computerized Factory: On the Shop Floor 136
   Flexible Manufacturing Systems
   Robots
   Management Information Systems
   Total Operations Planning System (TOPS)

6. Computers Off the Shop Floor: The Wider Context 217
   Computer-Aided Design (CAD)
   Computers and Managerial Structure
   The Global Factory

7. Computers as Strikebreakers 247

8. "A Technology Bill of Rights" 264

Notes 279

Index 297

ACKNOWLEDGMENTS

THE WRITING of this book was a long journey into many communities, workplaces, union halls, and universities. There are far too many people who contributed generously of their time and spirit to thank individually. I only hope that the final work merits their contributions. Among them were John Bennington, Mike Cooley, Adolfo Gilly, Dick Greenwood, Bennett Harrison, Seymour Melman, David Montgomery, David Noble, Bernard Roth, Merritt Roe Smith, and Joseph Weizenbaum. Brian Moriarity and Leo Marx read the manuscript and improved it considerably with their detailed comments. Some friends in Detroit provided intellectual insights, emotional support, and an understanding of the labor movement that was invaluable. I have much to thank Al Gardner, Russ Leone, Pete Kelly, Saul Wellman, Bob King, and Steve Kindred for.

Early work with Ivan Illich revealed some of the hidden ways in which technology shapes our lives and laid the groundwork for challenging the neutrality and inevitability of these changes. Without Marian Wood, my editor at Holt, there would have been no book. She saw the possibilities in some vague early ideas, and with enthusiasm, skill, friendship, and enormous patience saw this project through.

I am especially indebted to a close and long-lasting friendship with Alice and Harry Chester and Iris and Stan Ovshinsky. Their ideas and values had a decisive influence on me at an early age, and their support has always been there. My father, Max Shaiken, contributed a vision of courage under personal adversity. Finally, I would like to thank my wife, Bickie Manz, for her sacrifice and extraordinary encouragement during the writing of this book. Thanking her for this alone, however, would not come close to stating her real contribution. My ideas, goals, and achievements have been inspired as well as shared by her. Her love and support have made me a far richer and better person.

PREFACE

NEW TECHNOLOGY, based on computers and microelectronics, appears to be everywhere. What is being automated is not simply the typing of a letter with a word processor or the welding of a car body with a robot, but the collection, transfer, and control of information. As a result, few workplaces are exempt from the extraordinary changes now taking place. Journalists as well as typists, engineers as well as machinists, steelworkers as well as check-out clerks are all being affected. The technical potential of this transformation in production—higher productivity, better quality, and increased flexibility—is far-reaching and widely heralded. But, what about the social implications? How will work be restructured and the workplace reorganized? The prevalent view is that technological development is synonymous with progress. Most jobs will become more creative and satisfying. And if there are any negative effects, they are but a small price to pay for all the extraordinary improvements that are made possible.

Certainly, the technical potential is there for new systems to be designed in a way that both enriches life on the job and benefits society. But the potential to make work more creative and satisfying can be very far apart from the reality of the workplace. This book explores a significant and perhaps growing gap between promise and reality and the reasons for its existence. To do this requires stripping away the mystery that surrounds the process of technological development, exploring the managerial strategies that govern its use, and analyzing the experiences of workers and managers in the workplace. A central theme that emerges is that the design of new forms of automation reflects not only technical requirements but social decisions as well. The ways in which the workplace is revamped, therefore, are not a mandate or even simply a by-product of the technology but a result of conscious choices.

Ultimately, the way these choices are made and carried out has a direct and powerful impact on the lives of people at work. The changes now taking place are so extensive that they are redefining the meaning of work in the society as a whole. Take the issue of skill, a theme that runs throughout this book. Skill is a very difficult concept to define precisely. A useful definition might be "a creative response to uncertainty based on experience, ability, and the needs of the situation." Automation can be designed in a way that enhances human skill or it can be designed in a way that seeks to eliminate the human element entirely. When computerization is counterposed to human skill, the issue is not simply a "better" method of production but often a choice that reflects certain values and purposes. Is the purpose of minimizing skill to increase efficiency or is the central goal to extend managerial control over the process of production? This is a key question this book will explore.

The reorganization of production affects more than skill and skilled workers. In some cases, new forms of monitoring and control are embedded in the design of technology, further constraining workers on already tightly controlled jobs. Telephone operators in New York, for example, now have the time of their responses to customers monitored to the second as calls are being answered. The response time can be averaged by shift or by week, and operators monitored and ranked over time. In one office, the new twenty-seven-second average per call has resulted in a doubling of the number of calls handled in a day. Rather than provide for more creative jobs, automation in this instance has resulted in an electronic assembly line and undoubtedly higher levels of stress for the workers who are now electronically paced. The operators are not the only ones, however, who may experience negative effects from the system. The extra five seconds or so needed for a friendly personal remark to a customer or for a little additional aid, repeated over the course of a day, could seriously impair an operator's average; hence, the volume of calls handled may rise at the expense of the quality of the service.

For me, the use of technology to degrade work is hardly an abstract matter. I have spent an important part of my adult life as a skilled machinist. I learned the trade as an apprentice machine repairman in a large automobile factory. Later, I worked in the machine shop of a small research and development company, under far more creative circumstances. The contrast between these two very different work environments first sparked my interest in the different ways the same technology could be used. Subsequently, in the course of writing this book, I began to explore how the technology itself could be redesigned to achieve different social ends, how the values of the designers become embedded in the technology itself. My work experience is an important frame of reference informing this book. Underlying my analysis is the view that tragic costs result when technology is designed in a way that throttles rather than expands the enormous creative potential of human beings in the workplace. This book seeks to show that, given the potential of the technology, it is also an unnecessary cost.

The impact on workers of a pervasive and authoritarian development of technology can be personally devastating. Pete Kelly, president of United Auto Workers Local 160 and a skilled model maker, described the cost this way: "When people's work is diminished, their lives are demeaned." The workers interviewed in this book give ample testimony to the truth of his remark. Beyond the effects on the individual, there is also a political consequence of the transformation of work. Kelly, a militant trade unionist for over twenty-five years, is one of many who are concerned that the power of workers on the job is being undermined. To Kelly and other like-minded unionists the destruction of a worker's control over how a job is carried out, particularly a skilled worker, weakens the collective ability of workers to deal with their employer. The ultimate result is a weakening of the labor movement itself.

This book has yet another aspect. Increasingly, much attention has been focused on the competitive failure of U.S. industry. Some analysts have looked to the nature of production technology to explain this failure. The flexibility of computer technology is heralded as a technological answer to economic malaise. And the flexibility is credited with enriching work as well. Unfortunately, the possibility exists for introducing authoritarian principles in flexible as well as more traditional mass-production technologies. Under these circumstances, the extraordinary economic potential of these systems is not realized. The increased managerial control comes at the expense of worker initiative. The real issue is not simply about purchasing more automation but about how these new technologies are designed and deployed and why.

Ultimately, this book is about choices, both technical and social. It is about how these choices are made today and what their implications are for the present and the future. While all changes in the workplace are justified in the name of increasing productivity, this book argues that the issues of power and control are very much central. The choices affecting power in the workplace most immediately influence what happens on the job, but they ultimately affect the character of the society itself. A democratic society and an authoritarian technology are incompatible in the long run. If we are to choose one over the other, the development of technology in a democratic direction holds the promise of utilizing these extraordinary scientific achievements in a human way.

Introduction

THE FLEXIBILITY of microelectronics extends computer technology to every corner of manufacturing. On one automobile assembly line, for example, 98 percent of over 3,000 welds on the car body are performed by swiftly moving robots, computer-controlled mechanical arms. In an aerospace machine shop, the complex contours of a jet-engine support are carved out by the whirring cutter of a numerically controlled machine tool, itself directed by a computer that is simultaneously supervising over a hundred other machines. In a diesel engine factory, bearings are removed from one of over 4,000 identical bins by a cart that glides silently down 100 yards of darkened aisle and lifts its forks three stories into the air. In an engineering office, the structural properties of a part are tested by an engineer sitting at a cathode-ray tube—a TV-like screen—in Cologne, Germany, and then the finished design is instantaneously transferred to a computer in Dearborn, Michigan.

Each of these operations represents an extraordinary

change from conventional methods and an impressive technical achievement. The most far-reaching development of all, however, is the linking together of these separate processes into a highly integrated manufacturing system, the computerized factory. The transformation is often called computer-integrated manufacturing (CIM), or sometimes by the longer title computer-aided design and computer-aided manufacturing (CAD/CAM). Whatever it's called, the result is a revolution in the way things are made and the way work is organized. The technical basis for this revolution rests on the information processing capabilities of computers and the ability of microelectronics to bring computer power directly to the point of production whether that happens to be a machine tool or a word processor.

In the last several years, the capabilities of computers and microelectronics have soared while prices have plummeted. During the 1970s, for example, the cost of computer power declined an average of 50 percent every two and a half years.[1] This trend has contributed to the explosive growth of industries that manufacture new highly flexible forms of automation based on these technologies. Look at the production of robots. The revenues of robot producers have soared 50 percent annually between 1978 and 1981.[2] Or, consider the manufacture of computer-aided design equipment, a marriage of computer technology to the design process. The leading firms in this area have been growing at over 30 percent a year.[3] The recession in the early 1980s, of course, has dampened the market for automated equipment but this was often a case of deferred rather than cancelled purchases. (In fact, the 19 percent growth in robot sales in 1981 would have been highly desirable for most industries.) Overall, General Electric, which has moved heavily into this area, predicts that by 1991 the market for all forms of factory automation could exceed $29 billion a year.[4]

The workplaces already using these new technologies span

the gamut of manufacturing: industries from automobiles to machine tools, production volumes from millions of parts to one-of-a-kind prototype work, and factory types from five-million-square-foot transmission plants to storefront machine shops. General Electric, for example, as if to make its prediction on automation growth a self-fulfilling prophecy, spent over $2 billion to modernize its own factories in 1980 alone.[5] And, despite its battered condition after over four years of weak sales, the auto industry plans to spend $80 billion in a five-year program to bring out a new generation of small cars. By 1985, 255 domestic auto plants will be rebuilt utilizing the latest in computer technology, and billions more will be spent to convert thousands of supplier factories.[6]

This far-reaching technological changeover raises some pivotal questions that affect the entire society. Since any advanced industrial economy is heavily dependent on manufacturing, changes of this significance in the technology of production are bound to have repercussions far beyond the factory walls. For example, will these new technologies spur economic growth and the creation of new jobs or will they be introduced in a way that decimates employment? Will automation free workers from soul-destroying work or will it lay the basis for subordinating people to machines in a new and degrading way? Will computerization be used to rebuild the nation's industrial base or will it facilitate the ability of multinational corporations to operate independently of any given country?

In this book, I argue that the answers to these vital questions are not predetermined and depend on more than the technical capabilities of computers and microelectronics. The technology provides choices; what is selected depends on who does the choosing and with what purpose in mind. The following chapters explore the nature of the technological choices now being made, the ways in which they affect life on the job, and the changes they are spurring in the structure of the workplace. Looking only at the impact of technology, however, is too narrow and static a focus. It implies that the nature of the changes taking place are fixed and the consequences unavoidable. Instead, I want to uncover the social and economic forces that mold automation itself, and probe the other possibilities for structuring work that might be available.

Computers and microelectronics present a dazzling array of alternatives in the workplace: more worker autonomy or greater managerial authority, more skills or fewer skills, increased hierarchy or more democratic decision-making. In a market economy, the selection and assemblage of these electronic building blocks into a production system is largely the prerogative of management. Even in industries where there are strong unions, a clause inevitably exists in the contract that cedes the right to run the business to management. Take paragraph 8 of the UAW-GM agreement. It states:

> . . . the products to be manufactured, the location of plants, the schedules of production, the methods, processes and means of manufacturing are solely and exclusively the responsibility of the corporation.[7]

Management's central criterion for production is maximizing profitability. This theoretically leads to a search for production methods capable of producing the most output with a minimum of inputs, whether the input is materials, energy, or labor. In meeting this goal, some of the design choices are largely technical in nature. The scheduling necessary, for example, to assure that a load of steel arrives at the same time that a part is due to be machined is basically a question of coordination. Other choices are social. They affect what people do in the workplace. Workers, however, may have very different ideas than managers about what should take place in production, and the way these conflicts are resolved ultimately involves power.

For managers, power can be extended through the design and deployment of technology. A central theme of this book is that, driven by considerations of profitability, management's development of new machines and systems reflects the desire to increase control over production, over the activities of workers as well as over the movement of materials. This propels the design of automation in a direction that seeks to minimize the disruptive potential of workers. There are two ways to do this. One is to automate the job entirely; this is not always technically possible or economically feasible. The other is to reduce the input of the worker on the job. This calls for designs that reduce skill requirements, transfer decision-making off the shop floor, and exert tighter control over the workers who remain. The result can be more boring and stressful work in a more tightly controlled work environment.

Developing automation in this direction carries with it some high long-term costs for society. In fact, there is a fundamental contradiction between the potential of computerization to enrich working life and increase productivity and the development of the technology in the pursuit of authoritarian social goals. The moral cost is that people's lives become diminished through degrading their work. The productivity loss stems from the inability of systems that reduce the input of workers to fully utilize the skill, talent, experience, and creativity that only human beings can provide. Moreover, in seeking to bypass human input at almost any price, new systems often achieve a breathtaking complexity that is prone to breakdown and consequently requires even greater human input. Ironically, the drive to eliminate any dependence on workers who operate machines can result in a new and greater dependence on those who repair the equipment.

The productivity loss is frequently hidden because it is difficult to quantify. The output of new machines is compared to the production of the conventional machines they replace, which often results in spectacular improvements. Part of the improvement, however, stems from the fact that the new machines are more capital-intensive. A more meaningful comparison is the actual output of the new technology compared to what its potential output might be with design changes that more fully utilize the human capabilities available. Looked at this way, the drive for control winds up as a brake rather than a spur to productivity—a truly ironic twist since the design of the technology is justified by its ability to increase efficiency.

In most ways, automation based on computers and microelectronics forms a continuum with previous technologies. The latest form of computer chips, for example, is based on development work that goes back decades and on a body of scientific knowledge that has evolved over centuries. In other ways, however, this technology is qualitatively different from what has come before. Given its extraordinary capabilities, it could aptly be termed superautomation. There are three key differences in this technology: its flexibility, scope of application, and rate of technical change.

Consider first the flexibility of computer-based automation. When the term automation was first coined in the late 1940s, reportedly by a Ford Motor Company executive, it referred to transfer devices such as those that shuttle engine blocks between machines without human aid in automobile engine plants. Skill was built into the equipment in the form of steel cams to guide machine movements and fixtures to hold parts. In order to modify a part, however, the entire production process had to be rebuilt, calling forth welding torches, the remachining of parts, and plenty of skilled workers. Often, the price of the modification approached the original cost of the equipment.

In contrast, computer automation is characterized by its

flexibility. Programmed instructions adapt general-purpose machines to specific applications. If there is to be a change in the product, then new instructions to the machine are virtually all that is necessary. A robot, for example, might be used in the welding of the body of a large car one year and in the production of a subcompact the following year.

This flexibility leads to the second difference, the scope of application. The "hardness" of previous forms of automation limited their use to mass production where the high volume could justify the expense of the equipment. Computer automation, on the other hand, is economically viable for a few pieces as well as for several million. Moreover, the sophistication of the control systems allows for the automation of many types of jobs that previously could only be done by humans. The result is a technology that is simultaneously revolutionizing the office, the assembly line, the supermarket, the engineering department, the hospital, and the loading dock.

The third difference is the rate of technical change. In the past, the design of automation evolved slowly over years if not decades. For example, an engine line brought into an auto plant in 1964 would probably only be marginally different from one introduced in 1954. The development of computer technology, however, has been proceeding at an exponential rate. New levels of capability are often achieved in a matter of months rather than years.

Taken together, the properties of computer technology lay the basis for an unprecedented restructuring of the organization of production. The technology is used to integrate the various parts of the factory, speed the transfer of information, and remove the limits of time and space from the direction of a corporation. Managers gain the ability to mold production to their social and technical purposes whether on the shop floor of a single plant or throughout the operations of a multinational company.

Computers and microelectronics play the role of the central nervous system in the factory. Information pours out of computer terminals located on time clocks, machine tools, inspection stations, design terminals, and warehouses. There is a vertical integration from the designer in the engineering office to the machine tool on the shop floor. The designer sketches a part on a cathode-ray tube, and when the design is finalized, a set of instructions is automatically produced to guide the machine. In other words, the designer "talks" directly to the machine. Further, the technology makes possible a horizontal integration on the shop floor, linking together a wide range of production activities. Computer systems not only control the operation of machine tools, but track raw materials coming into the shop, inventory completed parts, monitor robots on the assembly line, and schedule production. And these different systems are designed to be able to interact: Machines "talk" to machines.

The information-gathering potential of the technology is impressive. Imagine the logistical problems of building a jet aircraft. Thousands of complex and expensive parts have to be machined in the right order and at the right time to be put into hundreds of intricate subassemblies, which in turn become part of the finished plane. To accomplish this at the sprawling McDonnell Douglas plant in St. Louis, 439 terminals in 41 buildings pour 100,000 transactions a day into a central computer.[8] In manufacturing alone, 25,000 different work orders are processed at any one time.

In the latest systems, the collection and transmission of data occurs as the event is taking place, literally at the speed of light. In the past, reports of shop floor activities were transmitted by verbal command and written memo, and except for the most pressing emergencies, communication routinely took place after the fact. Information more or less drifted up to the desired level. Now production problems are relayed instantaneously to the designated point in the production hierarchy, and routine information is compiled on a much more frequent basis. If the parts produced on a machine begin to drift out of preset limits, for example, a computer instructs the machine to make the proper corrections or, if the problem is more serious, the maintenance office is signaled. When managers need information, they can retrieve it instantly from the computer rather than wait for a weekly or even monthly report.

The same information-gathering ability that expands the horizons of the manager, however, can also be used to monitor more closely what workers are doing, weaving a net of control around their activities. The key issue is not the ability of computers to provide greater access to information, but *who* defines the data to be collected, who has access to it, and for *what* purposes it is used.

Telecommunications combined with computers remove the geographic limits from management. A transnational company is able to maintain the advantages of centralized control over managerial decisions in far-flung subsidiaries without sacrificing the benefits of decentralized operations. Sitting at computer terminals, engineers or managers can view the identical data at the same time from different areas of the plant, the country, or even the world. Basic policy decisions can be made at world headquarters and then refined as engineers throughout the company's operations interact with the same body of information. When sitting at computer terminals, these engineers are, in effect, in the same room for the purposes of planning and communication even if one happens to be in Paris, another in Detroit, and perhaps a third in São Paulo. Walking away from the terminal, the engineer is at the point where production takes place and therefore is immediately able to receive the kind of feedback that only comes from being on the scene.

Some of the more potent capabilities of superautomation are based on the ability to tie separate computerized machines and systems into larger networks. The integrated nature of these networks is mirrored by the corporate activities of the firms that are manufacturing automated equipment. General Electric is an excellent example. In April 1981, GE unveiled its master plan to strengthen its position in every area of factory automation. The plan involved heavy research and development expenditures, large capital commitments to internal product lines such as computerized machine controls, and a series of outside acquisitions. The corporation announced a $58 million expansion of its R&D center in Schenectady, New York; the construction of a $55 million microelectronics center in Raleigh, North Carolina; and the formation of a new industrial electronics business group. Then the company unveiled some of its outside acquisitions: $235 million to acquire Intersil, one of the largest producers of integrated circuits, a pivotal building block for automated systems; $170 million to buy Calma, the third largest producer of computer-aided design equipment; and an undisclosed amount to acquire Automation Systems, a robot organization to be used as the nucleus for a major thrust in this field. The emphasis in this entire endeavor is on fitting these various systems together, and all of this corporate activity represents only a beginning.[9]

In pursuing this analysis of superautomation, it is important to separate the technical possibilities of computerization from the social purposes that now influence its design and deployment. There may be an enormous difference between what the technology is capable of and the particular way it is currently being used. A computer's ability to collect data while an event is taking place, for example, does not have to lead to surveillance of workers. Unfortunately, technical possibilities and social purposes do not come in neatly labeled packages but are often intertwined in a complex and rapidly changing environment. To unravel them and see their relation, the analysis must be anchored in the realities of the workplace: the real world of whirring cutters, electronic controls, managerial directives, and worker activities. It might be simpler to theorize about the potential of automation without this taxing technical detail, but the analysis would fail to take account of the interaction of production systems with people that constitutes the work environment. Moreover, the development of technology is a dynamic process that changes in response to its environment. What happens on the shop floor today, both the cutting of metal and the activities of workers, influences the ways in which technology will be designed in the future.

Let me provide a road map of the way I will develop the analysis. Chapter 2 begins with a look at the machine shop, a pivotal industrial area that will be the setting for much of the subsequent discussions. Among the milling machines and lathes, it examines the ways in which the structure of the modern workplace has its origins in "scientific management," a nineteenth-century attempt at rationalizing and controlling the labor process, and the ways in which skilled workers today retain considerable authority on the job and the "moral code" that influences their behavior.

Chapter 3 probes some themes that influence the way new production systems are designed. In particular, it looks at managerial control as a criterion of design, minimizing human input as a pervasive engineering philosophy, and the schism between those who design machines and those who operate them as a factor in automation's development. In addition, it questions the extent to which the current direction of technological change blocks other paths of development.

Chapter 4 is a case study of numerical control (NC), the computerization of skilled machining. NC is the basis for a far-reaching restructuring of small- and medium-batch manufacturing, where parts are made in groups ranging from one-of-a-kind to several thousand rather than in volumes of hundreds of thousands, which is characteristic of mass production. These low-volume production runs, which were impossible to fully automate in the past, comprise a surprising 75 percent of manufacturing's share of the gross national product. NC was the first computer technology on the shop floor, reaching its current level of sophistication after almost three decades of development and use; it thus affords an opportunity to observe a mature technology in operation, one that in many ways may be a model for the way more recent computer technologies will affect other occupations. This chapter examines how the criteria of design, such as the attempt to bypass human input, play themselves out in a manufacturing setting. Not only does the analysis look at the way in which NC affects workers but also the way in which workers can affect the operation of NC.

I go beyond NC in chapter 5 to analyze the operation of the computerized factory as a whole. This chapter looks at the overall changes computers and microelectronics make possible in the organization of production and then explores a number of specific machines and systems in depth. It analyzes flexible manufacturing systems (an attempt at the automatic factory based on NC), looks at the use of robots, and studies the ways in which authority can be embedded in the design of management information systems (MIS), a technology of information gathering.

I leave the shop floor in the next two chapters. Chapter 6 analyzes the role of computers in design, the structure of management, and the globalization of production. Many of the conflicts about skill and power that exist on the shop floor are replicated on the level of design and managerial organization. This chapter also shows how computers and telecommunications pave the way for the "world factory." Chapter 7 probes the ways in which this technology can tip the balance of power in labor-management relations.

In chapter 8 I sum up the challenges automation poses

for the society in general and labor in particular. The chapter explores the strategies various unions have used to cope with technological change, and then poses some alternatives for democratizing technology.

Before I begin this exploration, my model for analyzing technological change ought to be made explicit. Ultimately, I argue that the workplace is shaped by the interaction of workers, managers, and technology at the point of production. Therefore, there is no simple cause-and-effect relationship between the potential of new technology to transform work and the final shape of the workplace. Although I focus on technology as a factor of production, a number of other potent forces also come into play, such as the type of work organization, the reliability of new machines and systems, the response of workers, and other managerial goals.

Consider the effect of work organization on technology. The same machine, say a computer-controlled lathe, might be used one way by an authoritarian management and very differently by a management that encourages employee participation. In the first case, the worker will likely be excluded from any decision-making, while in the second there may be an effort to obtain worker input on the job. Nonetheless, the design of the technology determines the framework in which these different styles of management operate. If the design of the lathe seeks to minimize human input, then there are preset limits to the skills and autonomy the worker will be able to exercise, regardless of the way work is organized. As a result, attempts to involve workers in decision-making may give only the illusion rather than the reality of power unless the shape of the technology itself is open to discussion. In other words, the style of management may encourage free choice after the technology has already limited the number of choices available.

It would be a mistake, however, to assume that designing

technology to minimize skills or limit human input and fully accomplishing these purposes are the same thing. If a machine's reliability is questionable or if the manufacturing process is unusually varied and complex, a need for substantial worker input may arise regardless of the design goals. This input strongly influences the level of skills actually required and the amount of leverage workers retain. Therefore, the impact of new technology must be analyzed in the unpredictable environment of the shop floor, with attention paid to the specifics of the workplace.

The worker's response is another important determinant of how new technology affects the work environment. On the shop floor or through a union, formally or informally, workers may push computerization in a much different direction than what management intends. Union pressure, for example, might result in a highly skilled and highly paid operator running a computer-controlled lathe, whether or not this is technically necessary. Once the skilled worker is there, more decisions can be made at the machine because the capability exists to address unusual problems that inevitably come up. Or, the union might win gains such as job rotation so that whatever skill level is actually required to operate the lathe will only be a part of the skills the worker needs to carry out a variety of tasks.

The drive for increased managerial authority is often associated with subdividing work. The ultimate example is a worker repeating the same job over and over again on the assembly line, easily monitored and paced by the speed of the line. Paradoxically, increased managerial control sometimes stems from the flexibility of deploying workers freely among a wide variety of tasks, thereby widening the content of their jobs. The automakers, for example, have been strenuously pushing to eliminate dozens of maintenance classifications such as electricians and machine repair and pipe fitters, seeking instead one or two broad categories of skilled workers who can handle whatever job comes up. Detroit obviously feels that this flexibility will cut the total demand for manpower and increase the utilization of the workers who are on the job. The price for this is sacrificing some of the authority that comes from subdividing work. The skilled workers, for their part, are struggling to maintain the classifications in order to protect their jobs and preserve their control over the work process. The automakers may ultimately be able to win both flexibility and authority by gaining the right to deploy workers as they see fit and also by using new technologies to simplify the work tasks involved.

In the final analysis, the design of technology is not only a question of machines and systems but of power and political choice. Computers and microelectronics lay the basis for highly productive manufacturing systems that fully utilize the capabilities humans have to offer and, consequently, enhance the work environment. A more participatory technology, however, requires a broad public awareness and democratic involvement in defining the goals for designing automation. Moreover, a major shift in the balance of power in the workplace would have to take place, giving those individuals and organizations affected by change—workers and unions—a voice in the nature of the change. Otherwise, as we will see, technological development reflects managerial purpose alone.

The Machine Shop

WALK INTO A MACHINE SHOP. As more and more jobs are concentrated in offices and in the service sector, it is an unfamiliar place for an increasing number of people. Upon entering, you immediately notice a number of massive gray machines, strange-looking shapes covered with an endless number of protruding cranks, handles, levers, and cams. These are the mills, lathes, drill presses, and grinders that are the heart of the shop. Their steel machine spindles and cast-iron work tables look polished and precise. Proudly cast into their heavy frames are the unfamiliar names of machine-tool manufacturers: LeBlond or Bridgeport or Giddings and Lewis.

You notice it is noisy. Metal is being ground, sawed, hammered, bent, and cut; compressed air is blowing chips out of the paths of cutters; heavy machine motors are in gear. It is smelly. Cutting oils, lubricants, solvents, and grease are exuding a semi-rancid odor that nonetheless has a fragrant quality to many machinists. By the standards of the office, it is dirty. Metal filings and long, curled steel chips are scattered over the machines, workbenches, and floor. It is, however, well lit with bare fluorescent fixtures hanging from the ceiling every few feet. To the novice, it all looks somewhat intimidating.

These humble surroundings have been the core of any manufacturing-based economy. It is in the machine shop that rods of steel and blocks of aluminum are transformed into the gears, cams, crankshafts, and other precision parts that are vital for industrial production. Anyone who runs one of these machines is a machinist, and within this general category exist many subdivisions. The United States Department of Labor lists four broad groups: machine tool operators who produce thousands of parts on specialized machines; set-up workers who ready these machines for production; all-around machinists who operate a variety of general-purpose machine tools; and tool and diemakers who produce the precision fixtures that hold parts while they are being machined and the metal forms that stamp sheet metal.[1]

Machinists work in remarkably different settings. The cavernous machine shop in the sprawling McDonnell Douglas St. Louis plant, for example, covers 750,000 square feet and is filled with towering two-story-high machine tools whose value can exceed $1 million. In contrast, a typical small machine shop near Detroit might occupy a few hundred square feet of unused warehouse space with a capital stock of one or two six-foot-high Bridgeport milling machines whose value is about $5,000 each.

The highly skilled all-around machinists and tool and diemakers—about half of all machinists—are the workers upon whom computerization is having the greatest impact. The complexity of the product and the low production runs, from a few hundred aircraft parts to one-of-a-kind diemaking, have made it difficult to automate this kind of work. As a result, small-batch machining has depended heavily on the skills of the machinists involved, which places them at a pivotal point in the production process and gives them substantial power on the shop floor. Accordingly, they tend to be among the most stable and cohesive work groups and among the most combative in defense of their rights.

It is the machinist who converts an engineer's conception, whether in the form of a drawing or a verbal description, into a finished part. This involves far more than good coordination and knowledge of which lever to pull on the machine. The skill is one of the brain as well as the hand, and it is learned on the job, transmitted from journeyman to apprentice and mastered only after years of experience.

The operation of a milling machine exemplifies the complexity of a machinist's job. Before any metal is ever cut some careful planning is required. The machinist determines the machine setup, plots the order of operations, and selects the speeds and feeds at which the machine will run. A block of material—say, steel—is then clamped to the machine table, the cutter is loaded and positioned, and the machine is turned on. Once the part touches the rapidly rotating cutter, an operator with years of experience is needed to spot potential problems and to react correctly when they do arise. A slight change in the color of the chip may mean the entire part will warp; a small difference in the sound of the machine could mean a poor finish; a mild chatter of the cutting tool might result in a scrapped part.

After all attempts to eliminate the experiential base of machining and to rationalize the requisite skill are made, an indefinable "feel" remains. One of the best machinists I ever knew worked in a small back-alley shop in Detroit making, among other things, precision copper flanges for high-technology vacuum equipment. Leo ran his lathe slower than the book said and fed the machine faster than the book said, put his elbow and some of his weight on the tool post to get a little more stability, clamped down on his cigar, and used his fifty-nine-cent paintbrush to apply a homemade mixture of kerosene and oil. The result was a perfect part. Leo had more pity than contempt for people who machined by the book, more or less the opinion of an artist for someone who paints by the numbers. No one told Leo to run the machine faster or not to lean on the tool post—not unless they wanted to take the responsibility for a scrapped part. Leo took that responsibility and thus had considerable freedom and control over his work environment. He loved making things, and pride in his craft reinforced his independence. At night or on the weekend, his children came into the shop to watch him build something or to make a few simple parts themselves.

Leo, like other machinists, plays a critical although unacknowledged role in the design process. This aspect of the trade appears in few job descriptions and is seldom if ever written about in engineering texts. But without it, production would grind to a halt. A familiar sight in most shops is an engineer walking in with a stack of blueprints to ask the worker if a particular job is feasible. The machinist carefully studies the prints, looks at the engineer, and says, "Well, it can be tried like this, but it will never work." Grabbing a pencil, the machinist marks up the print and, in effect, redesigns the job based on years of experience of what will work.

The price of this input into design is a reliance on the judgment and cooperation of the machinist. When workers feel harassed, they often begin producing parts exactly "according to the print." This "refusal to redesign" can be even more disruptive than a strike. In a strike, production is halted, but "work to rule" can produce mountains of scrap. Take the case of a machine repair shop in the auto industry. Supervisors frequently bring rough sketches or even broken parts from production machines to the machinists with little more instructions than "make one like this." The parts range from half-ton crankshafts for stamping presses to small custom-made bolts for the assembly line. Because hundreds of thousands of dollars of equipment might be idle waiting for the part, time is critical. When management initiated a campaign to strictly enforce lunch periods and wash-up time, the judgment of some machinists began to fade. At about this time a foreman dashed up to the shop with a "hot" job that was needed to keep the production machines operating. The job involved using a lathe to drill a 1¼-inch diameter hole through the center of a 4-inch diameter rod of solid steel, a simple task for an experienced machinist. Anxious to get the job done quickly, the foreman insisted that the machinist run the lathe at a high rate of speed and plunge the large drill through the part. Under normal circumstances the machinist would have tried to talk the foreman out of this approach but now was only too happy to oblige what were, after all, direct orders. The part not only turned out to be scrap, but part of the lathe turned blue from the friction generated by the high speed. The disciplinary campaign was short-lived.

SCIENTIFIC MANAGEMENT

The power skilled workers are capable of exercising on the shop floor is obviously incompatible with total managerial control over production. Attempts to limit this power, however, have had a turbulent history. One of the most far-reaching endeavors began in the last part of the nineteenth century at a time when craftsmen wielded considerable power in metalworking. Frequently, a skilled machinist became a subcontractor to a larger manufacturer, agreeing to deliver a certain number of parts in a specified period, usually a year. The way in which the parts were made was left to the discretion of the skilled worker. He hired, fired, and deployed his work force, effectively designing his own production process. The owners provided the setting for production: floor space, machinery, tools, and materials. At first, the skilled worker even paid his workers directly but later the actual payments were processed through the firm's financial department.[2]

The "inside contracting" method of management had considerable advantages and disadvantages for manufacturers. One advantage was that supervision was no longer the firm's problem. Skilled workers familiar with the production process provided clear direction to workers of their own choosing. The development and implementation of complex pay schemes became largely unnecessary. Weighing against these benefits, however, were a number of liabilities. The system provided management with only a dim idea of the exact cost of materials and labor, both in individual departments and in the firm as a whole. The coordination of production was largely left to informal cooperation among subcontractors, which could break down when conflicting priorities arose. Finally, this method allowed workers on the shop floor to retain substantial power. Ultimately, management had little precise knowledge, let alone control, over what was happening in production.

The rapidly developing industrial base soon burst the bounds on the "inside contracting" system. The emergence of large capital-intensive companies, the drive for competitive efficiency, the development of new technologies all played their part. In particular, the economic collapse of the 1870s underscored the limits of the old system. The prolonged depression brought with it excess capacity in metalworking and focused the attention of owners and managers on work organization. Men such as Henry R. Towne, the major stockholder of Eaton, Yale, and Towne, Captain Henry Metcalfe, the former superintendent of several federal arsenals, Oberlin Smith, the chief engineer of a New Jersey machine-tool company, all began looking at issues of accounting, coordination, and control.

These problems of the organization and improvement of management became central topics at the 1886 meeting of the recently formed American Society of Mechanical Engineers. At the meeting was a relatively obscure middle manager from the Midvale Steel Company, Frederick W. Taylor.

While not the first to look at these issues, Taylor developed ideas and methods whose impact resonated through the manufacturing community and, ultimately, the industrial world. A shrewd strategist as well as an effective popularizer, Taylor overshadowed most of his contemporaries who were also seeking new ways to structure production. He became the leading theorist and chief spokesman of what he eventually termed "scientific management." Taylor sought to convert the abstract right of management to "control" labor into a specific direction for every aspect of work. While many of the specifics of his system have long been forgotten, Taylor was nonetheless responsible for much of the managerial philosophy that underlies the organization of work today. In fact some of Taylor's goals in extending managerial authority anticipated the drives that now influence computerization.

Obsessed with rules and systems from childhood, Taylor proved to be a suitable prophet for the reorganization of industry on the threshold of mass production. His approach was to define and monopolize the knowledge of how to do work, and then return it to the shop floor in minutely divided and easily controlled tasks. The problem, according to Taylor, was that the knowledge workers possessed effectively allowed them to run the shop.

> This mass of rule-of-thumb or traditional knowledge may be said to be the principal asset or possession of every tradesman . . . foremen and superintendents know, better than anyone else, that their own knowledge and personal skill falls far short of the combined knowledge and dexterity of all the workmen under them.[3]

To define and monopolize the knowledge of the workplace, Taylor felt that "it becomes the duty and also the pleasure of those who are engaged in the management not only to develop laws to replace rule-of-thumb, but also to teach impartially all of the workmen who are under them the quickest ways of working."[4] Independent knowledge in the hands of the worker laid the basis for malingering or "soldiering" on the job. Taylor maintained that there was both "natural soldiering" and "systematic soldiering." The first stemmed from "the natural instinct and tendency of men to take it easy," and could be corrected by a manager who could either elicit cooperation or enforce obedience. Systematic soldiering, however, was a collective phenomenon, rooted in the mutual cooperation of workers to secure common standards. It compounded the problems of "natural soldiering" since "the better men gradually but surely slow down their gait to that of the poorest and least efficient." Taylor sought to cure this problem through redesigning work, thus eliminating the basis for soldiering by making production workers as interchangeable as the parts they were producing and skilled workers as limited and controlled as the technology would allow. Added to this would be the proper incentives that would make systematic soldiering unprofitable and therefore unattractive.

At the core of Taylor's system was time-and-motion study. While the idea wasn't new, Taylor claimed originality for one aspect of his approach: "a careful study of the time in which work *ought* to be done" rather than a historical record of how long it actually took. To carry this out, Taylor reduced jobs to their elemental parts, discarded inefficient movements, and then timed and recorded the necessary motions as they were done by the most skilled workers. The considerable guesswork that remained was buried in the seeming exactness of the extrapolations and in the apparent precision of the formulas. Taylor felt that this method would offer final and irrefutable "proof" to the worker of how the job should be done, transforming issues of power and authority into questions of communications.

Taylor expended considerable energy investigating the art of metal cutting itself, conducting upwards of 30,000 recorded experiments and many other inquiries of which no records were kept and using almost 800,000 pounds of steel and iron. Once obtained, the knowledge had to be monopolized in order to insure that management and not the worker would be in a position to control key manufacturing operations. Taylor justified this by claiming that ordinary machinists would not be able to understand the complex nature of his discoveries.

> The art of cutting metals involves a true science of no small magnitude . . . so intricate that it is impossible for any machinist who is suited to running a lathe year in and year out either to understand it or to work according to its laws without the help of men who have made this their specialty.[5]

Since anyone capable of running a lathe had already mastered this "true science" well enough to earn a living at cutting metal, many machinists no doubt felt differently about their ability to utilize the information from these new discoveries. Nonetheless, Taylor was insistent that machinists not be given the necessary information to make their own production decisions. Slide rules that one of his associates developed to determine feeds and speeds for the machines could not be left at the lathe "to be banged about by the machinist. They must be used by a man with reasonably clean hands, and at a table or desk, and this man must write his instructions as to speed, feed, depth of cut, etc., and send them to the machinist well in advance of the time that the work is to be done."[6]

Once the knowledge was obtained, it was to be centered in a planning department, which was to become the brain of the industrial enterprise, the center for coordination and control. The planning department would determine the capacity of the plant to produce and match this to the orders received.

Output standards would be "scientifically" set through job analysis and time-and-motion studies. Based on the orders that had been placed, the planning department would schedule the long-term work flow and also determine the daily work plan for each department and even for each worker. The tasks were to be sent to the shop floor in minutely divided jobs, carefully defined and easily measured. A worker was to receive written instructions a day in advance; these would describe what he was to do and how long he had to do it. There was to be a continuous monitoring of "the cost of all items manufactured with complete expense analysis and complete monthly comparative cost and expense exhibits." Even personnel functions such as hiring and firing would be set here. The orders of the planning department would be carried out not by supervisors who were generalists but by teams of eight functional foremen each of whom would have specific areas of responsibility, from setting the feeds and speeds of machines to being the shop disciplinarian. In effect, each worker would report to eight bosses instead of one. One can only guess at the problems of coordination this would have entailed.

The whole system was fueled by money. Taylor developed a complicated though psychologically crude "differential piece rate" system. Workers who failed to achieve the "scientifically" determined time received a lower rate per piece, and those who exceeded became eligible for an extra bonus 30 to 100 percent above the going rate. The full possibilities of the system would only be achieved when "almost all of the machines in the shop are run by men who are of smaller calibre and attainments, and who are therefore cheaper than those required under the old system."[7]

At the beginning of this century, a number of factories were organized on the principles Taylor had set forth. A fundamental roadblock, however, impeded the full implementation of the system: It involved the subordination of corporate management as well as labor to the engineers in the planning department. Nonetheless, some of Taylor's contributions—among them, the planning department—proved to be immensely influential; other ideas, such as functional foremen, didn't get very far.

Nowhere has the Taylor-like rationalization of the workplace been carried further than in the auto industry. Henry Ford's production methods have become virtually synonymous with the minute subdivision of work and time-and-motion study. In the heyday of the model T, when the tin black car was being purchased even faster than it could be produced and when Ford engineers were jamming more and more machines into the Highland Park plant, the vast factory provided a human laboratory to unleash Taylor-like methods on the work force. In their classic study of the Highland Park plant, Arnold and Faurote stated that the company preferred production machinists "who have no theories of correct surface speeds for metal finishing, [but] will simply do what they are told, over and over again, from bell-time to bell-time."[8] In order to oversee the system, Ford's machine shop alone required over 510 overseers of one type or another, each of whom had the authority to fire a worker. On the assembly line, the nature of the job not only was standardized but the worker was physically tied to the "scientifically" determined pace by the movement of the product along the line.

In the auto industry and elsewhere, however, there proved to be two critical limits to Taylorism's effectiveness—one social and the other technical. The social limit stems from the distaste workers quickly develop for highly routinized and regimented jobs. The technical limit arises from the impossibility of fully eliminating skill and control from many occupations through the organization of work alone.

The social fallout from Taylorism is absenteeism, high turnover rates, shoddy workmanship, and even sabotage. An early warning of this occurred with the introduction of the assembly line at the Ford Highland Park plant. The line brought with it labor unrest and the necessity of constantly hiring large numbers of workers to replace those who quit in disgust. Even the sources of the vastly increased productivity of those first heady days at the Ford plant have recently come under closer scrutiny. As William Abernathy puts it in his book *The Productivity Dilemma*, "The moving assembly line was certainly important in its own right, but its contribution may have been overstated. Productivity gains came from a series of changes, like those in length of the workday and in wages."[9]

As a result, there have been many more sophisticated attempts at motivating workers and eliciting their cooperation since Taylor. In the 1920s, for example, Elton Mayo, a Harvard sociologist, was instrumental in developing the Human Relations school of management.[10] Sharing Taylor's concern about workers consciously producing less than they were capable of, Mayo sought to combat this restriction of output with a more sympathetic management style and employee counseling. In more recent times, there has been a widespread movement toward more employee participation in workplace decision-making. A wide variety of programs are grouped together under the general term "quality of work life." Whatever its merits in creating a better environment on the job, however, in virtually all cases, management retains the right to design, deploy, and use technology as it wishes. Under these circumstances, while workers seemingly receive more power to organize work, they lack real control over the technology, which provides the context in which work is organized. Rather than expand worker control, the end result could be one of more fully adjusting workers to managerial deployment of technology.

The technical limits to Taylorism stem from the complexities of the manufacturing process. In metalworking, in particular, it is often easier to provide guidelines than the definitive solutions that Taylor sought. In trying to determine the "one best way" to cut metal, for example, Taylor himself worked with twelve variables and countless permutations of these variables. He investigated metallurgical questions such as the quality of the metal to be cut and the chemical composition of the cutting tool; he analyzed workpiece geometry such as the diameter of the work and the depth of the cut; and he looked at what happened during the cut itself such as the thickness of the chip and the elasticity of the workpiece and the cutting tool. Given all these variables, there will be a direct relation between the volume of parts produced and the effectiveness of Taylor's methods. If a million of the same type of crankshaft are being machined, for example, it is technically possible and economically feasible to fine tune more accurately the metal cutting and to determine what the worker should do on the job. Moreover, in mass production hardened steel cams build skill into the machines, providing repeatability and production reliability with less dependence on the worker. Without high volume, however, important aspects of skill and control remain with the worker. If only a few hundred copies of the same part are being produced, it is economically unfeasible to define accurately the parameters of the cut, let alone exactly what the worker should do.

THE MORAL CODE

The control that a machinist retains over how work is carried out is by itself only potential power on the shop floor. The way that power is realized also depends on a strong sense of shared values that inform most activities on the job. Four central concerns stand out for skilled workers: pride of craftsmanship, independence on the job, a sensitivity to changes in skill and job content, and a strong sense of collective action. Together these themes comprise a moral code not essentially different from that of nineteenth-century craftsmen and one that can lead to considerable workplace independence. In describing skilled workers in Taylor's time, David Montgomery might well be speaking of many machinists today:

> Technical knowledge acquired on the job was embedded in a mutualistic ethical code, also acquired on the job, and together these attributes provided skilled workers with considerable autonomy at their work and powers of resistance to the wishes of their employers.[11]

These attributes are very much in evidence among the tool and diemakers in Ford Motor Company's enormous Rouge complex. Straddling the Rouge River ten miles southwest of Detroit, the plant is a monument to Henry Ford's dream of making the Ford Motor Company a vertically integrated producer, controlling all aspects of the business from the mine to the finished automobile. Mammoth Ford-owned ships float iron ore, coal, and limestone into the center of the complex, which converts these raw materials into gleaming Mustangs that are driven off the assembly line at the rate of one a minute. Spread over the compound are coke ovens, a steel mill, a stamping plant, a powerhouse, a glass plant, an engine plant, a frame plant, and a tool and die building. The plant now employs about 16,000 workers, down from 27,000 in the late 1970s, of whom 2,000 or so are skilled metal workers. Of these, 800 work in the tool and die building, the largest die shop in the auto industry. The rest of the machinists and mechanics are scattered through the various factories in the Rouge, working in smaller shops or on the production floor. All the metalworkers are represented by the politically powerful tool and die unit of United Auto Workers Local 600, the bargaining agent for all hourly workers in the Rouge.

Consider what a diemaker does. Starting with blocks of steel and some drawings, a diemaker builds a contoured steel form capable of stamping thousands of sheet metal parts, say automobile fenders. Some diemakers operate machine tools such as mills and lathes, some work on the bench filing and fitting the die, and others work on the production floor repairing and modifying the dies after they are put into the stamping presses. The work requires considerable coordination and communication among different groups of diemakers to insure that all the parts come together at the right time.

Although diemaking can be hazardous, heavy, and dirty, there is a great deal of pride in the trade. Take Russ Leone, a burly thirty-eight-year-old who has been a diemaker since the mid-1960s. Leone, who served an apprenticeship at a small shop in the Detroit area, came to the Rouge in 1970. In 1978 he was elected to the full-time job of union committeeman and is now vice-president of the unit. A tough and sophisticated unionist, Leone enjoys being a diemaker:

> Building something that works from nothing is the most satisfying part. You've done all these operations yourself to put it together. You can't just want to get by. You have to care about what you're doing and take a lot of pride in it.[12]

This pride can result in some strange twists between the stereotypes of managers concerned about quality and workers trying to slide by. A harried foreman behind on the manufacture of a critical die might be tempted to send it to production before a diemaker thinks it is ready. According to Leone, "If a diemaker doesn't think it's good enough, he doesn't like anyone else to say it's good enough."

In the case of some diemakers, releasing a die early could be a serious mistake. Once old Louie began working on a die, for example, he felt that it ceased to be the property of the company. On more than one occasion, he responded to what he felt was an incorrect management decision by having a tantrum, an activity involving hollering, gesticulating, and throwing a part or two. To the embarrassment of many a foreman, Louie didn't care who was around when these rages began. Some foremen would plead with Louie's coworkers to talk to him about restraining these outbursts. If Louie had something to say about quality, could he please say it to his supervisors when they were alone and not to the entire plant management team. Given the intensity of his response, Louie was obviously an extreme but not an isolated case. In a rather concentrated way, he expressed feelings about pride in craft that most diemakers share.

This pride leads to a desire for independence on the job. Viewing themselves as highly qualified professionals, diemakers resent unsolicited advice and resist outside control over their work, preferring to be told only what to build, not how to build it. As *American Machinist* magazine puts it, "Diemakers themselves are an independent lot who put their trust in their own skills and are generally reluctant to let go of the personal control they have over the finished product."[13]

Al Gardner understands this need for independence. A diemaker himself for over thirty years, Gardner was born in England and learned his trade at British Leyland. He has worked at the Rouge since the mid fifties and is now president of the tool and die unit. Widely respected in the unit as an intelligent, militant, and independent trade unionist, Gardner describes one source of the diemaker's autonomy: "You have to make the die on your own, and since you take responsibility for it, you don't want to be pushed around." The complexity of the job lays the basis for substantial power on the shop floor. Leone elaborates:

> Management depends on your knowledge. A foreman has thirty dies to build. Now there is no way he can build thirty. He might be able to build one but there is no way he can know what move to make next in every die. The best foreman is the one who seeks to do the least.

For the foreman who seeks to intervene actively, the dangers can be great. One diemaker recalls an incident in the stamping plant at the Rouge:

> We had an old foreman on production who was a real nice guy, a guy named Mike, and we had a pretty good relationship. When you work in the pressroom after awhile in a certain area, if a press goes bad or a die goes bad, you know what to do before you even get there, you've done it so often. The foreman will say, "I have a problem on such and such a press," and we say, "Fine, we'll go and take care of it." But, they got a new guy there, a new production foreman, and he started to lay the law down. He just kept harassing us. So one day he said to me, "I need the punch details sharpened on such and such a press," and I said, "Fine I'll do it at lunch time." Instead of doing them in place, which wouldn't take very long, we took them all down and shut the press down for three or four hours. He never gave us any trouble after that.

Diemakers, like most skilled workers, are extremely sensitive to what they perceive as the weakening or erosion of their trade. Turf is jealously guarded. In particular, skilled workers fear that increased management flexibility could undermine their power on the shop floor and ultimately result in job loss. Stemming from these concerns, thousands of grievances are filed every year challenging the company's blurring of the divisions between trades. Actions such as an electrician removing a bracket that is normally taken off by a diemaker are hotly contested. Other grievances dispute management's delegation of skilled work to production workers. For example, activities such as a machine operator performing minor repairs on a machine are sure to be swiftly challenged by skilled workers.

These seemingly picayune squabbles are, in reality, disputes over more fundamental questions of power and job security. The real issue is who will organize work on the shop floor. Work rules and past practices not only define and limit managerial rights, they codify relations of power at some point in the past. Skilled workers defend work rules because they lack the power to enforce a more flexible organization of work. These work rules, however, often assume a life of their own, remaining in force even when the underlying grievance has disappeared.

This sensitivity to the narrowing of skill or the undermining of craft manifests itself when diemakers vote to ratify or reject collective bargaining agreements. In the 1973 negotiations between Ford and the UAW, the company established the right to schedule production workers to do limited maintenance and repair when all skilled workers had been offered the overtime and refused. This apparently secondary part of the agreement precipitated an overwhelming rejection of the contract by skilled workers at Ford. The national vote was 6,000 for ratification, 20,000 for rejection. The tool and die unit voted the contract down by 1,000 to 76.

To implement their code of conduct, diemakers act collectively. All workers have leverage when they act together, particularly by walking out, but diemakers have an unusual number of tools at their disposal. They are capable of fine tuning their response to a managerial directive by speeding production up or slowing it down without resorting to more extreme tactics. When they do walk out they can delay the introduction of new models, which gives them devastating leverage. Some of the tactics diemakers use may appear petty and obstructionist. In some ways they are, but these activities frequently stem from the diemakers' inability to affect their immediate conditions on the shop floor in any other way. Frozen out of the decision-making process, the diemakers resort to the leverage that is available. Moreover, these tactics often come in response to petty and authoritarian managerial directives. Consequently, what may appear as constant hassling to management represents to the workers involved a defense of self-esteem, pride of craft, and decent conditions on the job.

Supervisors are aware of this potential power and frequently strike informal deals offering, say, to relax work rules in exchange for emergency production. One diemaker describes the process: "A foreman will come up to you and say, 'My ass is in a sling, help me out and I'll punch you out four hours early.' If the foreman has been a decent guy, the diemaker will do it."

Over time, however, diemakers work according to group norms that are set informally, though collectively. The code is formed and transmitted when diemakers meet informally on the job: over a cup of coffee, in the locker room, at the job lineup, or at lunch. Diemakers will ask each other what they have done already or what they have to do and perhaps compare this to the workload they know is in the shop and the level of urgency the respective jobs have. If someone gets out of line, a few comments generally will straighten things out. The same diemaker maintains: "You're not going to work in a way that makes the people around you look bad. They could always do the same to you. You know what is expected of you and you do it."

The development of these standards and most other forms of concerted action are strongly reinforced by the leader system. Leaders are diemakers who direct the activities of five or six other diemakers. Promotion to the position is based strictly on seniority. Oddly enough, some workers who were not the best of diemakers often become very good leaders. As a group, the leaders are quite respected. While they earn somewhat more than a regular diemaker, they remain hourly workers and union members. Their first loyalties are generally to other diemakers rather than to the company.

In the tool and die building, the seventy or so leaders are known for effectively coordinating their actions. George Tapocik, a leader and now a bargaining committeeman, claims, "The leaders are the kingpins in the building. They are a very close-knit group that functions together to build the dies and they work together on quite a few other things as well." Al Gardner echoes these sentiments. "Leaders run the tool and die building. They are very, very tight. You upset one leader, and you've upset them all." Jack Donovan became a leader in the tool and die building in 1965. A diemaker for thirty years, he was until his retirement in 1982 a popular and respected vice president of the unit. He describes some of the leaders' attitudes:

> The leaders are interested in their job and they talk to each other quite a bit. They're really good tradesmen and they like to build dies. And they want to see their trade upgraded. They don't want a part of anything that will downgrade it. The guy that knows his job real well doesn't want to see it changed or eliminated.

Much of the leaders' power rests on their pivotal role in organizing production. They decide who goes on what job, and how the job is built. Moreover, they coordinate the construction of the die with other leaders on the benches, on the machines, and in the tryout area. Donovan describes the nature of the job:

> The biggest decision is who's going to work what job. It makes a big difference who works on what job because if some guys hate to work on a type of die there is no use putting them there. It will take them twice as long to build and won't be worth much when it's done. After they are assigned, the diemaker pretty much knows how to do it, he just needs some reassuring now and then. And you have to lay out an overall plan as to how you're going to proceed. After that you may get with the machine people to make sure the details are coming through all right.

The union has learned how to use this power. According to Gardner,

> I've worked in the building since the fifties, and from what I've seen, anything that needs taking care of in there, the leaders are the ones to do it. If the union wants anything done in the building, we know that the leaders will do it. They are the people that stick together the most. The company is afraid of the leaders. They know that if the leaders are involved they've got a real problem. They know for a fact that if we tell those leaders something, or even if we don't tell them, that as long as they're planning to take some action, the whole building could come to a screeching halt.

Though the individuals change, through retirement or leaving the company, the basic code of conduct remains similar.

If, in fact, the leader does all of this, there doesn't appear to be very much left for the foreman to do. Donovan continues:

> The only thing a foreman is good for is making sure the diemakers don't punch out early or go to the john every morning to read the newspaper. And he is sometimes helpful if you need to set the priorities. But, as far as building the dies? I'll tell you, the times you need a foreman are few and far between.

This lack of involvement can lead to considerable frustration among the foremen, some of whom were leaders themselves. Donovan analyzes the causes of these frustrations: "I think what bugs the foremen is that they don't have any control, or at least very much. The leaders direct the men and the foremen are just standing off in the aisle, really."

Sometimes the foreman seeks to get overly involved in building the die to relieve these frustrations, according to some of the leaders. But the foreman lacks the intimate hour-to-hour contact that the leader has with the job. Donovan states the problem:

> The leader and the diemaker may have already looked at the job and agreed to do it a certain way. They'll say, "OK, we'll put this section in and we'll have to cut that one, do this, and we'll cut this corner, and that's the way it will work." And it will work. Then the foreman comes over. "How are we going to do this?" The leader explains what they said they were going to do. "Oh, no, no, let's not do that. Let's do this, do this, and do this." Well, he's made his point that he's the boss and some leaders will say, "Yes we will do this, we'll do anything you want." Some won't let it go by. Some you cannot take their power away. But others will just say, "Whatever you want." And they'll call the diemaker back over and say, "Here's the way we're going to do it now." And they do it and of course it's all screwed up.

Donovan proposes a rather unusual solution to the problem of frustrated supervisors and their consequent intervention in production:

> You'd be better off bringing in foremen that know nothing about diemaking. They should go out in front of the plant on Miller Road and hire foremen right off the street. Then let them come in and handle all the menial work, and let the leaders build the dies. Then the foremen could just get the hi-lo and track down parts, things you have to do. Right? Really, he'd be wielding some power then and wouldn't be so frustrated. The way the building runs the best is when the leaders run it.

The union organization on the shop floor can also be extremely effective in using the varied and complex nature of diemaking to win gains. One committeeman describes a unique form of psychological bargaining:

> On the bench it's not easy to tell if a guy is giving you a bad deal. Guys will be mad at their boss and they'll screw the boss, but they have a problem, the boss doesn't know he's getting screwed. The best tactic I can use, this will really shake them up, I'll go up to the general foreman, and I'll say the guys need coveralls or they won't work. All you have to do with the foreman is plant the seed, and they come unglued, because they're not sure. Then every time they see two guys standing together they get flustered, because they think the guys are screwing them.

There are a range of other tactics. In one plant, when management began rigidly enforcing discipline on breaks and wash-up times, all the templates—a critical diemaking tool—mysteriously disappeared. In another building at the Rouge, a superintendent sought to enforce shop discipline by docking the pay of thirty or forty workers who left the work area early. The committeeman responded by informing workers that the superintendent wanted to know they were there. After this, every time the manager came out on the shop floor, every worker would stop his job and begin banging the machinery with whatever was available. Not only was the clatter deafening, but production slumped in direct proportion to the number of trips the superintendent made to the shop floor.

The collective power of diemakers sometimes brings about some extraordinary changes in the way parts are produced. There is a vast reservoir of talent on the shop floor that normally is untapped but that management is compelled to seek out during a crisis in production. By examining these situations, alternative possibilities of organizing work that do utilize the skill and creativity available become apparent. In contrast, Taylorism appears as a fetter on productivity, a method of maintaining managerial command that carries a high price.

In one case, management brought a set of battery brackets into the dieroom on a Thursday afternoon and requested that the workers build new dies capable of stamping those brackets immediately. The existing dies, they were given to understand, had somehow been damaged and final production was endangered. The diemakers jumped at the challenge. They were being asked to design and build a series of dies in less time than it might normally take to process the paperwork on the design. They worked furiously over the weekend, building a set of blanking, forming, flange, and pierce dies from raw stock. The dies were ready for tryout on Monday. Before shipment, however, the diemakers learned that the dies were in fact being built to provide an alternate source for a part produced in a plant where the workers were on strike. Their immediate response was to refuse to try out the dies, let alone to stamp parts. When they thought it necessary, they put extraordinary energy and inventiveness into building the dies. Had they known the real purpose of the project, it is likely that a thousand problems would have arisen that would have considerably slowed if not completely halted work.

In another instance, management sought to enlist the aid of diemakers in design because of a crisis at the end of a new model run. The workers received prototype parts and were instructed to design dies to build these parts. The diemakers became so involved in the design of complex dies that they effectively created a mini-engineering department on the shop floor, squeezed between tool boxes, lay-out benches, and machine tools. The diemakers not only did rough design sketches but they translated these sketches into sophisticated mechanical drawings and set up a highly professional system to check these for errors. The diemakers involved were excited by this fuller participation in a production process that had reintegrated conception and execution to a remarkable degree. Moreover, they felt a justifiable pride in their ability to produce designs that many observers felt were superior to the ones the company contracted to professional design shops.

This engineering activity came to a halt when management told a diemaker he *had* to design a die. The diemaker refused. He was then told "design or go home" and he chose the latter alternative. The other diemakers, although pleased by the expansion of their duties, stuck by him and stopped work on the design.

Meanwhile, the union maintained that any design work done by diemakers ought to be considered a new classification and paid accordingly. The company refused, contending that similar design work had been done by diemakers in the past, particularly when there was trouble on a new model. While other incidents like this have no doubt taken place elsewhere, this case provides unique documentation because the union filed a grievance and fought its case through the entire grievance procedure to the final stage of an umpire hearing. The umpire, an outside party acceptable to both the company and the union, ruled in favor of the union, concluding that the diemakers in question were doing design work of such sophistication that a new classification was needed. The company responded by refusing to give diemakers formal design work.[14]

At the same time, the union filed another grievance, further illustrating the integral role of diemakers in the design process. The union was demanding a new classification for diemakers who reviewed the die drawings that came into the shop both from outside vendors and Ford's own engineering department, making the necessary corrections and design changes. Both sides agreed that diemakers had always reviewed part drawings. The disagreement was over the extent to which this was now taking place. The union maintained that since several diemakers were doing review full-time, their pay and classification ought to change. The company argued that this was only a temporary increase in workload rather than a new job. This time, the same umpire ruled in favor of the company. He emphasized the wide-ranging nature of a diemaker's job, stating that "review of die construction prints is one of the basic skills and duties of the tool-and-diemaker classification."[15]

At times, different parts of the moral code conflict. Pride of craft, for example, can supersede collective action or the other way around. In some departments at the Rouge it is traditional to give retiring supervisors an automatic cigar cutter, a miniature press made out of aluminum and brass with a little electric motor. At the touch of a button, a miniature die turns over and cuts the end off a cigar. Dozens of shop hours are painstakingly put into building the cutter. Shortly before the presentation of one cutter, a worker was fired for trying to carry some paper toweling out of the plant. Outraged at the contrast between management's tolerance of using shop time to build gifts for supervisors and the firing of a worker for petty theft, a furious committeeman grabbed the cigar cutter and was about to throw it into a bailer to be smashed. The diemaker who built the little press grabbed it back from the committeeman, and yelled, "You aren't touching that die. I spent hours building, polishing, and perfecting that thing. You'll have to throw me into the bailer first."

But some diemakers in the same department later displayed a different attitude. A cigar cutting machine for a particularly hated foreman was stored in the supervisor's office prior to a party to celebrate his transfer. At one o'clock in the afternoon on the day before the party, the machine vanished. Management began a frantic search, offering all kinds of inducements including paid days off. No one could be found, however, who knew anything about its whereabouts. The timing was hardly accidental since it was too late to build another machine in time for the party.

The code may be interpreted differently depending on some other factors as well. Traditions and customs on the job are often different in large plants where there are layers of supervisory control than in small shops where the owner may be on the next machine. Moreover, when the economy is faltering, workers in any shop might be most concerned about having a job at all and will temper their militancy accordingly. Finally, the overall work relations in the society affect the conduct of a worker on the job. Swiss workers in a factory that has never been on strike undoubtedly have different attitudes than British workers in a plant that is struck every other month.

Consider the differences in work culture between the large die rooms owned by the auto companies and the small shops that act as outside suppliers. In the past, the auto companies offered job security and benefits at the price of extensive supervision, while the small shops provided money and independence on the job with often shaky job security. Today the automakers enjoy less than absolute job security and the small shops often pay less than the big three, but the tradeoff between extensive supervision and independence still exists. Underlying these differences is the reality that making cars is the primary business of the auto companies, while the die itself is the principal product in the small shops. Ironically, the result of more independence in the small shops is often a much faster work pace.

The extensive bureaucracy of the large shop often conflicts with the independent nature of the diemakers. The layers of supervision—foremen, general foremen, assistant superintendents, superintendents, and the plant manager—appear to be a police function rather than an aid to production. As Al Gardner put it:

> When I started at Ford's, they said, "If you're on the job at starting time, and at lunch time, and at quitting time, you'll do all right." You can goof off all day long or you can work yourself to the bone, but if you're not there you'll be in trouble.

Dave Matthews, a diemaker at the Rouge, concurs:

> What I like to do is work hard for an hour and then take a break. My foreman doesn't like this. He thinks that if I stay on the machine I might do something by accident. Our supervisors are policemen, not skilled workers.

The attempt to maintain tight shop discipline often backfires. One group of diemakers worked furiously for a couple of days without breaks to help a foreman complete a rush job. When the job was done, the foreman bought everybody a cup of coffee and the diemakers sat down to enjoy it. At about this time, the general foreman wandered by, saw everyone drinking coffee, and wasted no time in chewing out the foreman and the workers involved for sloppy work habits.

In the outside job shops, the layers of supervision and many of the petty work rules are absent. The task is building the die. In the smaller shops, the owner is generally a diemaker who not only supervises the shop but frequently pitches in on the machines. If a machinist is not up to his standards, the worker might be fired the first day. The diemakers who remain have quite a bit of independence. In a tight labor market, the threat of leaving provides the worker with even more leverage.

If the dies are built on time, activities that would cause rapid dismissal in the large, more tightly run shops are tolerated. One diemaker, for example, works hard for three or four months, completes a project, and then goes on a drinking spree for two or three weeks. After that he returns to work ready for the next assignment. According to Al Gardner, who spent some time working in the job shops:

> These guys who work in the job shops were like individual businessmen, they were contracting their skills. They were in there for the money, they were selling their tal-

ents, and what they wanted was money and some freedom.

In the outside shops or in the Rouge, a diemaker acquires the moral code that informs behavior when learning the trade. The apprentice learns more than the operation of lathes, mills, grinders, and shapers. Working for four years with a variety of journeymen, the apprentice also becomes aware of the work attitudes and practices of the shop. On the job, the journeyman, not the foreman, is boss. The journeyman determines what the apprentice will be allowed to do and therefore how well the trade will be acquired. When experienced diemakers begin work in a new shop, they quickly learn the code that governs the shop's activities. If a new worker stays on the job to the quitting bell and the other machinists wash up fifteen minutes early, the new worker will receive a warning. If the warning is not heeded, the diemaker may have few people to talk to and little help when a tough job comes up.

For management, worker control on the shop floor is an opening in the circuit of manufacturing control. In order to command the operation of the machine, the activities of the machinist must first be controlled. Instead, a hundred years after Taylor, managerial decisions that conflict with the moral code trip the breaker.

A key limit to the replacement of worker initiative in small-batch machining for Taylor and his successors was the lack of a new production technology. As a result, Taylor was compelled to rely on work reorganization to gain his ends. While he developed better cutting tools, faster machines, and more efficient methods of power transmission, the technology did not exist to change the relation of the operator to the machine. In fact, a machinist who worked in a machine shop in 1880 would have little trouble getting used to a conventional machine shop today. With the advent of low-cost computer control, much of this will change.

Criteria of Design

COMPUTER-BASED AUTOMATION is not found in nature. These new machines and manufacturing systems are designed by human beings who have certain purposes in mind, both technical and social. To understand how these new technologies are used to restructure the workplace, let's first look at the social goals they are designed to meet. Two pervasive managerial purposes stand out: reducing the amount of direct labor and increasing control over the manufacturing process. This chapter will focus on the second aim and its reflection in the design of technologies to command and coordinate the use of tools, materials, and machines. More complete control of the workplace, of course, also requires tighter control over the activities of the workers who remain. This extension of managerial authority is reinforced by two other related factors: an engineering ideology stressing predictable and automatic operation at all costs, and the near total separation of the people who design machines from those who use them. The realization of these purposes, however, is limited by some constraints. One such constraint, for example, is "the state of the art"—the whole history of a particular kind of technology as it manifests itself today. Nonetheless, the nature of these social choices is pivotal in shaping new forms of technology.

The final report of the Machine Tool Task Force, the most comprehensive study of the U.S. metalworking industry ever undertaken, offers a revealing insight into the social criteria applied by engineers in designing new machines and systems. The two-and-a-half-year project was funded and guided by the U.S. Air Force and involved hundreds of experts from the machine tool industry, user industries such as the auto industry, research organizations, and universities. The only major group not represented were the workers who operate machine tools and the unions who represent them. The five-volume report is filled with technical insight and expert methods to make metalworking more productive. A recurrent theme emphasized by the authors, however, is that technology should be designed to minimize the need for skills and to diminish worker decision-making. It is assumed that increased productivity and profitability will flow from a work environment in which all variables are eliminated or controlled. Since human participation is certainly a variable, one of the central conclusions of the study is that machines should be designed to "reduce operator involvement." One way to achieve this is to "simplify controls to allow use of lower-skilled labor" and to "approach unmanned operation."[1] One of the long-term recommendations of the report, to be achieved in this decade, is that industry should:

> Reduce the skill levels required to operate or maintain certain machine tools (or to plan the manufacture of a part), an approach already practiced by some of the technically more advanced companies. This can be done by using more automation, and substituting computers for people in executing certain decisions or operations.[2]

Iron Age magazine, a respected metalworking weekly, also contends that the manufacturing of the future will require fewer skills and less worker input. The writers seem breathless in describing the possibilities.

> Gone are the skilled workers.
> Reduced to almost nothing are the roles played by semiskilled machine operators.
> Out the window is much of the conventional in metalworking machinery and many of the traditional concepts of how and where to use that equipment in the manufacturing process.
> Virtually eliminated is the human handling of components and workpieces.[3]

The entire issue of skill is really part of a larger topic: the role of technology in revamping the organization of work. This question is best explored by looking at a specific technology. Out of the countless possibilities, computerized machining or numerical control (NC) is a prime example. NC, which is discussed in detail in the next chapter, involves transferring the control of a machine from a skilled worker to a preprogrammed set of instructions. The person who prepares these instructions is generally a part programmer. One respected text, written to introduce engineers and managers to the benefits of NC technology, argues that tighter managerial control is the necessary corollary of a more minimal role for the machinist.

> To a great extent, computer and numerical controls were *designed to minimize* the number of processing decisions made on the shop floor. Such decisions, whether they are good or bad, are nearly always suboptimal. Since the machine operator is largely outside of the machine control loop, manufacturing by automatic controls makes

> *tighter management control both possible and imperative* [emphasis added].[4]

The officers of the Numerical Control Society, an organization representing several thousand engineers, managers, and professionals concerned with the computerization of production, concur with this view, maintaining that a central advantage of computerization is more "direct control."

> Important decisions that affect unit cost, delivery dates, and quality are, with NC, in the hands of managerial and professional personnel rather than machine operators. This provides the ability to estimate costs more accurately, promise delivery dates with greater assurance, and provide repeatability that keeps a customer happy.[5]

Another benefit they point to is "company-owned skill—NC tapes don't leave a firm for a better offer; some operators do."[6]

The importance of expanded authority is repeatedly stressed. Here is what a now-classic text, which analyzes and projects the impact of computers on manufacturing, has to say on the subject:

> A recent step, one which has received widespread attention in management, is management's ability to issue commands to the machine on the floor. "Management has at last regained control of the factory" is a sentence often heard. The first step toward this achievement was numerically controlled machine tools.[7]

The increase in managerial control at the expense of skill is viewed by some as the technological equivalent of Frederick W. Taylor's attempts to reorganize the workplace a century ago. While this is an overly simplified and one-dimensional view, it conveys the significance of an important dynamic.

Iron Age describes the way in which NC reflects the spirit of Taylorism:

> Numerical control is more than a means of controlling a machine. It is a system, a method of manufacturing. It embodies much of what the father of scientific management, Frederick Winslow Taylor, sought back in 1880 when he began his investigations into the art of cutting metal.

In case we have forgotten what Taylor set out to do, *Iron Age* reminds us:

> "Our original objective," Mr. Taylor wrote, "was that of taking the control of the machine shop out of the hands of the many workmen, and placing it completely in the hands of management."[8]

The capacity of technology to provide increased control over workers is a frequent topic at machine tool shows and technical conferences. In some cases, the development of computer technology is counterposed to the need to motivate workers. Viewing the creation of a new work ethic as compounding the problems of dealing with already cranky and high-strung skilled workers, some managers and engineers are searching for new ways to technically limit the leverage of machinists rather than for methods to coax workers into compliance. Russell A. Hedden, a former president of the National Machine Tool Builders Association and also the chief executive officer of the second largest machine tool builder in the United States, is a strong and frequent proponent of this view. He presented this analysis in the keynote address at the sixteenth annual meeting of the Numerical Control Society in March of 1979. The scene was the plush Airport Marriott Inn in Los Angeles with about 600 people in attendance. Hedden delivered his address in the familiar manner of technical meetings, reading a prepared text illustrated by slides designed to emphasize the talk's main points. In this case, the slides were cartoon caricatures. One slide showed two workers with hard hats and lunch buckets strolling along and chatting away. In the background, a sexually appealing young woman in a bathing suit was posing seductively, visibly upset at being ignored. The woman was wearing a beauty-pageant-type sash that said "MISS MOTIVATION." In the dimly lit auditorium, Hedden declared:

> While I am a proponent of realistic employee motivation techniques, I do believe it is high time we accept the posture that major productivity increases in the foreseeable future will result primarily from investment in improved equipment, particularly where numerical control is involved. Motivation of workers can no longer be relied upon as the fundamental stimulus for improving national productivity. . . . We must recognize the effect of this new work ethic on the economy, then we must counteract the effect by devising more high-technology machines. These machines—many of them using new NC concepts—will more than compensate for any human productivity lag.[9]

Hedden then thanked the numerical control professionals in attendance for removing control of the machining operation from the machinist:

> . . . let me say here that you in numerical control have done us a great service by placing the control of many machining operations in the domain of the process engineer.[10]

Sharing Hedden's view are many plant managers. Viewing themselves on the front lines of production, these managers prefer to face the myriad problems of manufacturing without the need of simultaneously confronting an independent work force. They are also very influential in the purchase of machine tools. The views of a manager of a medium-size Massey Ferguson gear and shaft plant in Detroit are typical. He feels hampered in his ability to effectively manage the plant by government affirmative action hiring regulations, strict seniority provisions for laying workers off, and a UAW local he perceives as wanting to challenge any decisions he does make. Numerical control is an oasis of managerial sanity in a world going slowly out of control. Sitting in his office overlooking a dilapidated industrial belt on the west side of Detroit, he comments on the advantages of NC:

> You no longer have the same work ethic or skill level in an employee due to government regulations and other factors. We need a machine that takes control out of the operator's hands at least 90 percent of the time, and instead we must put emphasis on management and process people and programmers rather than on a journeyman who knows everything.[11]

He views the largest single advantage of NC as the "ability to set up and run a part without the high skill level of the operation. When the machine has control, management has control."

This view is shared by Doug Keno, the industrial engineering manager in the Massey plant. Keno stated:

> We're going to NC because it puts the process in our hands. The parts are made the same way every time. The operator can't change anything, he must do it the way you say. Young people in particular aren't interested in listening so we lean towards NC.[12]

The need to design machines requiring fewer skills is generally justified by pointing to a serious shortage of skilled workers or at least skilled workers with the right attitudes. A 1981 report of the Defense Science Board, for example, sounded the alarm by quoting, among others, a National Tooling and Machinery Association prediction of a shortfall of 240,000 machinists by 1985.[13] A technological solution to this shortage is offered by George P. Sutton, a scientist at the Lawrence Livermore Laboratories and the project leader of the Machine Tool Task Force.

> Many users, faced with the problem of a less capable and skilled work force in the future, believe that one solution is to partially remove the machinist from the operation of the machine tools. By relying less on the man in the decision-making process in this operation, by further automation, and by going even more toward . . . simple operator controls, it is possible to lower the skill levels of the operators of machine tools and reduce their required numbers. As other sections of this MTTF survey indicate, it has fortunately been a direction of the technology to aim toward highly automated systems.[14]

A senior Bendix sales engineer stresses the "less capable" aspects of today's work force in a devastating though highly inaccurate portrait of computerized machine operators.

> The characteristics of the typical operator in today's factory vary widely. Generally speaking, he is the least skilled, least educated, most transient and least concerned individual of the disciplines involved in the operation of an NC machine. He also represents the one area that has the highest potential to cause a wreck and is least trained on the machine's proper operation.[15]

In response to this alleged state of affairs, "the manufacturing industry is favoring the purchase of machines with controls that require less operator attention to oversee the process."

This concern is repeated throughout the trade press in articles and advertisements. Typical is an article under the headline, "AUTOMATION BECOMES VITAL: A SHORTAGE OF SKILLED WORKERS IS FORCING FIRMS TO TURN TO COMPUTERIZED MACHINE TOOLS." The piece continues by saying, "The nation's tool and diemakers, most of them in their fifties and sixties, are being replaced by the computer's small fry cousins that run machine tools."[16] An article on automation at Lockheed-Georgia expresses similar sentiments.

> As an older workforce retires, man-years of skill and experience go with them. It is not possible to hire the experience that you lose when an employee retires. For some jobs, those which are unpleasant or strenuous, it is difficult to hire even inexperienced personnel.
>
> Therefore, Lockheed is trying to capture the skill of experienced personnel in technology. The increase in automation compensates for a declining workforce.[17]

Moog Hydra Point, a builder of computerized machine tools, ran a full-page ad in the trade press asking readers in a banner headline, "HAVING TROUBLE FINDING SKILLED MACHINISTS?" The answer, of course, "TRY A SKILLED MACHINE!"[18]

Whatever the shortage of skilled machinists might be, it did not magically appear but is a result of some past failures. Two important shortcomings have been a lack of adequate incentives and of training. Rather than addressing these complex underlying problems, computerization becomes part of a self-fulfilling prophecy: Since automation replaces skill, there is less incentive to train skilled workers. The need for automation then becomes even greater.

The desire to extend managerial authority in the workplace propels more than the design of machines on the shop floor. Computerization is also used to peel off human involvement from every part of the production process, from the designer in the engineering office to the worker on the shop floor. The process is dynamic. Not only are existing jobs transformed or even eliminated, but even those jobs created as a result of automation are automated as the technology improves.

One existing job that is being revamped is process planning. The process planner has the key role of determining how a part is made, selecting the processes, the tools, the machines, and the sequence of operations. There is of necessity a close relation between the planner and the machinist. When a new job comes in, for example, the planner might go down to the shop floor and chat with the machinist about how a similar job was routed a few weeks ago. Any self-respecting senior machinist has a legendary "black book" that records the problems encountered and the shortcuts discovered on previous jobs, usually in some indecipherable shorthand. That carefully constructed knowledge, however, is the property of the machinist, enriched over time, and very much part of the workers' control over production.

One of the central thrusts of computerizing process planning is for management to gain control of that knowledge, removing an important area of dependence on the worker. According to *American Machinist*:

> That old-time machinist might be a vanishing breed, but the wealth of experience contained in that black book is today being captured and stored, allowing shops to improve their productivity by gaining control of their processes.
>
> Tomorrow's version of the machinist's black book is likely to be the automated process-planning system. The

> legendary experience will be the process data collected by the computer and stored for use by the process planners on identical or similar parts.[19]

The idea, however, is not to remove skill from the machinist in order to give it to the process planner. Planners are themselves too arbitrary and subjective: "Process plans frequently reflect only stubborn commitment to the personal experience, preference, or even prejudice of the particular planner or, perhaps, a parochial view screened from alternative methods that may actually be readily available under the same roof."[20]

Consequently, the computerization of process planning goes well beyond Taylorism. Taylor sought merely to gain control of the shop by transferring "brain" work to the planning department. Now there is a drive to automate the planning department itself. One example of this is computer-aided process planning (CAPP), one of many commercially available software packages. The developers of this particular program praise its ability to remove skill from process planning as one of the system's leading advantages.

> The CAPP System provides a means of reducing the individual skill and experience generally necessary for a process planner to function productively. It is a vehicle for standardizing and optimizing production methods by removing individual decision-making from the process planning function as much as possible.[21]

Beyond the machine tool industry itself, an additional force pushing technological development is the military. The Air Force, for example, was pivotal in influencing the design and encouraging the diffusion of numerical control. This influence was based on considerable funding. Today, the Air Force plays an even broader role in technological development through the Integrated Computer-Aided Manufacturing

Program (ICAM). Formally established in 1977, the effort is funded at $100 million through fiscal 1984. It is supposed to provide seed money to advance the state-of-the-art of automation by engaging in projects that may be beyond the capability or interest of any individual firm.

One of ICAM's central aims is to define an overall "architecture of manufacturing" that will effectively be a blueprint for the automated factory. The program seeks to develop modular computerized subsystems that will be the building blocks of this larger effort. The prospectus describes these as follows:

> These subsystems are designed to computer-assist and tie together various phases of design, fabrication and distribution processes, and their associated management hierarchy, according to a prioritized master plan. At appropriate times these mutually-compatible modules will be combined, demonstrating a comprehensive control and management package which is capable of continual adjustment as production needs and state-of-the-art change.[22]

But, what goals should this "architecture" and these subsystems be designed to meet? The prospectus details one central emphasis.

> . . . the Air Force learned that managers' interests went beyond increased labor productivity. Industry managers qualified their interest in CAM (Computer Aided Manufacturing) concepts on the potentials of extent of return on investment (ROI), maintenance of competitive position, greater design flexibility, and greater management control. Of these, industry considered management control as having the greatest payoff potential in CAM.[23]

At the present time, the Air Force believes that "there is still too much human involvement with set-up time, raw stock selection and feeding, and product removal."[24] But, the specific goals for a pilot project, an integrated sheet metal center, will seek to remedy that. The project targets a 44 percent cut in the number of people and an eventual return on investment of 24 percent.[25]

Whatever short-run gains might accrue from minimizing or eliminating skill, this strategy could extract a high long-term cost. Skill and experience are a vital human repository of manufacturing knowledge. Once an entire generation loses them, they are very difficult if not impossible to restore. The individual skill of the worker is important but this is only part of the collective skill of the work force. When an especially tough job comes into the machine shop, workers share their experiences and talents in solving it. The result of these interactions is greater than the sum of its parts. Technological design that seeks to bypass skill also severs these interactions and the extraordinary insight that frequently comes from them.

In some cases, the application of computers to production creates new skills or even new occupations. This development is more a statement about the technical limits to total automation than a refutation of the design criteria discussed. In fact, one outcome of computerization could be a two-tiered workplace: a small number of creative jobs at the top and most other jobs with fewer skills and subject to new forms of electronic monitoring and control. David Rockefeller, hardly America's premier social critic, predicted that present forms of automation might lead to a "two-tier society, with satisfying and well-regarded work for some, while the rest are left to grapple for unskilled jobs."[26]

Control is not an end in itself but a route to maximizing profitability. At times, methods that increase managerial command over production conflict with other objectives such as reliability or producing parts quickly. As will be seen in the next chapter, the machining process is so complex that more worker input than management desires is often necessary for the machine to operate at all. The technical challenge is to secure that input while limiting the ability of the worker to pursue an independent course. Simply put, in machine design there is a tension between allowing flexibility and seeking more control.

The tradeoff between flexibility and control is partially governed by the relations between workers and managers in the plant and in the wider community. Relations of trust within a factory may make managers more disposed to cede some power over the job to workers on the shop floor. Moreover, long-term consensus and cooperation on a national level allow a somewhat freer experimentation with alternative designs for technology and shop organization. In Sweden, for example, more sophisticated employers have experimented with a number of approaches to the organization of work, among them the team assembly of engines at Saab Scania and the group assembly of the entire vehicle at Volvo's Kalmar plant. Concerning NC, the Swedish Employer's Confederation describes one company that has found:

> the person operating the machine often is better placed to judge the machine's potentials [sic] and limitations than anyone else. Therefore, it is in the company's interest to give workers a large degree of influence in programming the machines.[27]

This experimentation in the organization of work, however, is not meant to alter the balance of power between workers and managers. And, the potential for disruption exists in even the most congenial of atmospheres, providing further limits to the amount of authority management is willing to give up.

Underlying the specific criteria of design for new machines and systems is a pervasive engineering ideology that views human participation in production as inherently "unscientific." This view is so widespread that it usually goes unstated. But, it serves to propel technological development toward the elimination of human input. Engineers tend to perceive a fully automated process as the ultimate solution to factors, human and otherwise, outside direct control.

An alternative approach does not begin by counterposing science to human experience but rather with the premise that the kind of knowledge underlying skill is still imperfectly understood. A 1981 report on new technology from the Council for Science and Society in Britain elaborates on this theme.

> The most awkward aspect of skill, to the modern, rational understanding . . . is that it always contains some element which cannot be fully analyzed or explained. The doctor, in arriving at this diagnosis, may use rules which he has explicitly learned or formulated, but he will also use others which he is not aware of knowing, and which he has never made explicit. This has been discovered as an empirical fact by the "knowledge engineers." For example . . . experience has also taught us that much of this knowledge is private to the expert, not because he is unwilling to share publicly how he performs, but because he is unable. He knows more than he is aware of knowing.[28]

It is possible to define and codify some of this implicit knowledge and the results can be extremely beneficial. But, the end product is a logical set of rules that, no matter how sophisticated, necessarily lacks the required judgment only human beings can bring to a situation. In many cases, the most effective route to solving a problem comes out of an interaction of experience, intuition, and ordered thought. What follows, then, is the need to develop from this conception a technology that enhances human judgment rather than a technology that aims, as a matter of principle, to eliminate it.

To appreciate the difference, let us reconsider the prevailing viewpoint. Dr. Bernard Chern, the former program manager of applied physics and production research of the National Science Foundation, maintains, "Today, large parts of manufacturing are still experience-based. They do not rest on a coherent body of analytically derived knowledge having wide applicability. This program seeks to provide such an underlying knowledge base."[29] Based on this assumption, the next step is transferring intelligence to the machine. Chern continues, "The first industrial revolution involved the transfer of skills from man to machine, and much has been accomplished in this area. The second industrial revolution, which is in its infancy, involves the transfer of intelligence from man to machine."[30]

Gideon Halevi, the author of *The Role of Computers in Manufacturing*, expresses the way in which this philosophy affects his own research:

> In the past few years, my research has been devoted to . . . the engineering phases of manufacturing . . . in an attempt to replace "art" by "science," that is, to replace intuition by computation, while turning skill and experience into formulas.[31]

And the Machine Tool Task Force continues this theme:

> Today the designer relies on his own or the machinist's experience; sometimes their combined intuition faces up to the uncertainty by deciding to try an experiment and see what results . . . fully integrated, computer-aided design-manufacturing systems will be paced by objective data bases that replace today's experiential and intuitive decision-making by people.[32]

The search for objective laws to guide manufacturing is the logical sequel to Taylor's search for "the one best way" to perform any task or organize any operation, only now computers are involved. The approach is based on the assumption that the countless variables of the manufacturing process can be so fully delineated and so completely quantified that the decision-making process can proceed by formula. If there is one optimal solution to every production problem, then once that solution is determined, any human effort to "improve" it is bound to be counterproductive. In certain applications, this may be the case. But "the one best way" as a universal procedure dangerously overreaches and becomes the search for a technological Holy Grail: a production process without human input.

In practice, the difference between these two approaches can be striking. A small midwestern town, for example, decided to computerize its emergency telephone service, which summons the police and firemen in an emergency situation. As a start, the record of calls for the previous several years was analyzed and codified. Then, a computer program was developed that would automatically assign emergency responses based on the severity of the call and the availability of vehicles and personnel. An act of violence in progress would rate a very high priority and divert police cars from, say, an unarmed theft if need be. Finally, the emergency operators would remain on the job to answer calls, feed information to the computer, and inform the caller about the action being taken. All the decision-making, however, would reside with the computer. The operators apparently were retained because they were less expensive and more familiar to the user than currently available computerized voice simulation.

After a trial period, however, this attempt to eliminate human input had to be abandoned. The operators did not like the downgrading of their jobs, but this objection alone undoubtedly would have been overriden. More important to those who set up the system, it was quickly realized that there is a real need for human judgment in answering an emergency call. An experienced operator is capable of detecting varying levels of urgency from the sound of a caller's voice and can ask probing questions that may result in a much different impression than might be had by plugging the initial facts into a computer.

In this case, the computer program was ultimately kept as a guide but the actual decision was made by the operator. New techniques exist to insert some of these nuances into the program itself, thus improving its usefulness as a guide, but the question remains: Is it desirable to seek the total eclipse of the operator from the decision-making process? And if the operator is removed, what happens when a truly unusual situation arises? Will the person, accustomed only to answering the phone and plugging the information into the computer, be capable of the appropriate response?

The report of the Council for Science and Society underscores the mistaken way in which "science" is often counterposed to human experience. Discussing Taylor's research in metalcutting in the nineteenth century, the report noted that one of his more significant discoveries was that cutting tools made from high-speed steel were capable of being run seven times faster than those of carbon steel, the material most tools were then made of. In order to use these new tools in the most efficient way, Taylor sought to standardize the tool shapes and developed a slide rule to calculate the optimum speed to run the machine.

Taylor now faced a choice. One direction would have involved going to the machinists, explaining his discoveries, and teaching them his methods for determining how fast to run high-speed steel. This would have converted what was then an implicit part of their skill, the determination of machine feeds and speeds, into an explicit series of guidelines. Instead of losing skills, however, this new knowledge would have created the basis for a more complete command of metalcutting.

Instead, Taylor made the selection of feeds and speeds a function of the planning department. The "scientific" discovery became a vehicle for removing skill and downgrading work. In the process, an unnecessary tension was created between "reason and science on the one hand and rule of thumb or tradition or guesswork on the other."[33] As the CSS report notes, this conflict was not new even in the time of Taylor:

> Thus science, when embodied in technology, becomes the enemy of skill, and of those who earn their living by selling their skill. This is not a new thought. Already in 1835 it appealed to Ure: ". . . when capital enlists science in her service, the refractory hand of labour will always be taught docility."[34]

An important element of human experience is removed from the design process by the near total separation of the engineers who design new machines from the people who will have to operate them. Since all technological choice involves a series of tradeoffs, this separation biases design decisions in the direction of sacrificing the quality of life on the job. If an equal degree of productivity can be obtained from a design that has considerable job satisfaction or from one that is alienating, at the very best there is no pressure to go in either direction. In fact, the divorce of design from operation makes the choice unknowable to the designer. A further consequence of this separation is that it becomes more difficult to incorporate the knowledge and experience of the worker in the design process. A worker who sees an operation day in and day out would be an invaluable aid to an engineer developing a new system.

This separation is not new. Without examining the historical record, it seems probable that the large ships that once plied the Mediterranean, propelled by legions of slaves in the galley, were designed by people who had no expectation of doing any rowing. (Otherwise, the development of more efficient sails might have been speeded up considerably.) More recently, when Henry Ford introduced the assembly line, he did so with the assurance of not having to spend any time working on one. In machine design today, workers are almost entirely excluded from the development process. And engineers are enveloped by managerial goals beginning with their technical education in the university and continuing through their entire working career. Whatever the merit of these goals, the aspirations of workers are only indirectly considered, to the extent they are considered at all. Engineers do not win promotions and universities do not gain research grants on the basis of developing systems that are the most satisfying to the workers who have to use them. Nonetheless, these ultimately might be the most productive systems.

Would machines be designed differently if those who used them had a more active part in the process? The answer undoubtedly is yes. The alternatives that haven't been posed are the flip side of the criteria of design that we have been discussing. What would a technology look like that stresses the enhancing of human skill or the increase in shop floor autonomy? In what university or research lab is development work in this direction being carried out?[35] At issue is not the choice between conventional methods and computerization, but the design and development of computerized processes in a way that would fully utilize the potential of human beings in the workplace. The intervention in die design by diemakers at the Rouge plant, discussed in the last chapter, provides a compelling example. Imagine the development of a work organization and technology that would be capable of incorporating this extraordinary input on a regular basis. Computerized systems could easily be designed in a way that facilitates locating design decisions on the shop floor.

A fascinating case of the tradeoffs involved in the design process comes from the Israeli military. The Mirkava tank, built in Israel, is considered one of the safest in the world. When it came to cost-benefit choices, its designers chose not to sacrifice very much in the way of safety, particularly the ability of the crew to escape in an emergency. In the 1973 Six-Day War, Israel had sustained heavy tank losses in the Sinai. Because the tank designers knew that in any war, they or their sons or other relatives were likely to be in a tank, they attached central importance to safety. Had the designers been further removed from the operators of the tank, they might have chosen to increase firepower at the expense of the tank crews' safety.

Once a certain design path is selected it can be very difficult to alter directions. For example, a community decides to provide better transportation for its citizens. It has two identically priced options. The first involves a system of underground subways, the second is based on a massive network of multi-lane highways. Say the second option is chosen. Once the highways are built, the enormous fixed investment weighs against any change. Even if the advantages of the subway became apparent, the choice is no longer equal. In the case of the computerization of the workplace, the widespread deployment of new systems is only beginning today. Defining design goals that move in a human-centered direction becomes vital if the option is not to be foreclosed.

Numerical Control: A Case Study

IN THE LAST CHAPTER, I stepped back from the machines and systems themselves to highlight the social goals that influence the design of the technology. Now I return to the point of production to probe the ways in which the development and use of a specific technology, numerical control, embodies these goals. This is an important place to begin for three reasons: first, numerical control is a fundamental building block of the computerized factory; second, skilled machining is one of the most complex processes to automate, which implies that other tasks would most likely present fewer problems; and finally, NC represents three decades of shop floor experience and technological refinement.

Numerical control is so radical a departure from conventional machine shop practice that the United States National Commission on Technology, Automation, and Economic Progress describes its importance as "probably the most significant development since the introduction of the moving assembly line."[1] It is not, however, a new method of cutting metal but a means of information processing and machine control. In fact, metal is cut in the same way as on a conventional machine, using the same types of drills and cutters. The difference is that an NC machine is controlled by precoded information while a conventional machine is guided by the machinist. Gone are the wheels, levers, cranks, and dogs that the machinist set by hand. In their place are automatic systems responding to electronic pulses.

With the machine now controlled by symbolic command, a number of options exist technically as to where and how the machine instructions should be prepared. One option, historically the first to be used and now by far the most prevalent, is to split in two what the machinist formerly did. The planning decisions are made away from the shop floor by a part programmer who determines what the machine will do and then translates this information into a form that can be read by the machine. The operating responsibilities remain with the machinist, who is largely reduced to making adjustments if something unexpected happens or stopping the machine if an accident occurs. The worker becomes a monitor rather than a participant in the production process.

The system was originally called numerical control because groups of numbers determined the operation of the machine tool. For the more complex programming languages of today, however, a more appropriate name might be symbolic control, but as Joseph Harrington puts it, a machinery buyer "would be deeply prejudiced against any machine in which the name of the system implied that he would have only token control of the machine."[2]

In this configuration, where planning is severed from operating, numerical control reflects the social criteria of design already discussed: fewer skills, less worker input, and increased managerial control. Once a program is tested, a high level of skill is permanently embodied in it. That skill ceases to be the property of the worker and therefore no longer can be used at the discretion of the machinist. As already noted, skill is not only a matter of job content or work satisfaction but the basis for important leverage in the workplace. A process that threatens skill undermines that leverage and lays the basis for a considerable transfer of power off the shop floor.

Numerical control, however, does more than affect the social relations of production. It also makes possible a significant increase in productivity and a remarkable expansion of the capabilities of metalworking. Is it, then, a technical advance or a means of social control or both? More importantly for this analysis, is the social control an inevitable byproduct of the technical advance or are there other alternatives? In the last chapter, I sought to show that designing technology to extend managerial authority is only one out of many possibilities. Now, in the case of numerical control, I will also argue that selecting this authoritarian alternative throttles the productive capacity of computers and microelectronics. For one thing, computers tie the various parts of production together, from design to cutting metal. Seeking to isolate the machinist from any real input in this highly integrated system cuts the entire operation off from vital shop floor feedback. For another, when machines are designed to require few skills, few skills may be available when they are needed. And, a further cost is the destruction of creative and meaningful work, an ironic development given the capacity of computers and microelectronics to enrich work.

The operation of NC on the shop floor, however, often does not fully reflect these criteria of design. There is a popular vision of NC as the centerpiece of a fully automatic self-correcting factory. As a number of examples in this chapter show, this is not yet the case. Critical technical and economic limits make it impossible to eliminate skills completely in small-batch production. Instead, varying levels of human intervention are needed and these give workers an opening to re-exert some control on the job. The result is a use of technology with more human skill and control than the designers may have intended. Am I arguing then that NC really doesn't work? No, it *does* work and it is being successfully developed in a more automatic direction. But, this evolution toward greater automaticity is warped by the social goals that inform the design of NC. There are really two different ways to move toward more automatic production. One way continues to emphasize the value of human judgment. Current levels of technology do not fully eliminate unpredictability, according to this view, and therefore the ingenuity of humans is the most effective corrective to this uncertainty. So while specific parts of the process become more automatic, humans retain control of the system. The second way seeks to leap toward fully automatic operation. Since the goal is eliminating human input, the most efficient blend of human and physical resources most likely will not result. In practice, reality pulls the system back to a less automatic operation: Things don't work unless there is "unofficial" human input.

In addition, other forces press in the direction of preserving human decision-making. In some cases, unions have successfully sought a greater worker role in the operation of NC. In other cases, sophisticated managements, especially in small shops specializing in intricate prototype work, have felt that efficient production requires skilled workers. Under these circumstances, the NC machine is viewed as a tool for the machinist, who often does a considerable amount of programming, rather than as a vehicle to control the worker. Although this may be the exception rather than the rule, it offers a glimpse of alternative production possibilities.

Nonetheless, the design of the technology determines the context in which these forces play themselves out. If a technology is designed to limit skill, it can sometimes be used in other ways but the range of possibilities is far more restricted. And the next generation of numerical control could very well succeed in eliminating some of the input that gives workers leverage today. The development of those possibilities that enhance skill are likely to be ignored.

This chapter will investigate four interrelated areas: the

technical and economic advantages of numerical control, the use of the technology on the shop floor, programming and computer numerical control (CNC), and the impact on the machinist. In the first section, I survey several ways in which computerization expands the capabilities of metalworking and a few managerial limits to its effective use. In the second section, I examine the ways in which numerical control affects life on the job, some of the limits to removing skill fully, managerial ambivalence as to how skilled a worker to retain on the machine, and some of the more subtle effects on the organization of work. In the third section, I first discuss the central question of where to locate production decisions, on the shop floor or in the engineering office? Then I look at an important recent technological development, computer numerical control, which adds considerable flexibility to the operation of NC and opens up some far-reaching possibilities for increasing worker control. I then explore the factors influencing whether this new potential is realized. Finally, in the last section I inquire into the perceptions of machinists concerning this technological revolution and the factors influencing these perceptions.

But, should all of this be of any concern to someone who isn't a machinist or who isn't directly concerned with what is happening in metalworking? The answer is unequivocally yes. The experience of numerical control today represents a case study about the role of technology in the conflict between worker control and managerial authority in the workplace. As we will see later, the same issues raised by NC are present throughout the factory and in a wide range of other occupations.

TECHNICAL AND ECONOMIC ADVANTAGES

The technical and economic advantages of numerical control that accrue to management are potentially impressive. Three principal gains are the creation of economies characteristic of mass production in low-volume manufacturing, the ability to produce complex parts that could not be made any other way, and the flexibility that makes possible rapid and varied new product development.

The most significant economic benefit is less direct labor for a given volume of work. The technology has the potential to be three to five times more productive than conventional machines, and this factor is often multiplied even further because machinists are assigned to tend more than one machine. When NC is linked to larger computer systems, its productivity edge can soar to twenty to one or even higher. Ultimately, NC leads toward the continuous flow of mass production rather than the stop-and-go of stand-alone machines and batch manufacture. The entire production process is streamlined with less setup, fewer machines, a more intensive use of machines, and less inventory. One important saving is not having to build expensive jigs and fixtures for each separate part. Rather than using these elaborate steel positioning mechanisms, the part is clamped to the machine table and the cutter automatically glides through the necessary angles and contours under the program's control. Jobs that formerly required a variety of machine tools such as drill presses and mills are now all done on the same NC machine, called a machining center, saving more set-up time, material handling, and, particularly, labor. This, and the fact that more operations are preplanned, lays the basis for faster throughput: less time from the order of a part to its delivery. Finally, there is a reduction in inventories. Rather than run larger than necessary batches and storing the parts for future use, the program is saved and run again when needed.

In addition to the direct economic benefits, NC can machine the most intricate of parts. Historically, numerical control emerged out of the search for a better way to produce the complex airfoil of a helicopter blade and was driven in its early days by the need for improved methods to manufacture involved parts for military aircraft. Today virtually any part that can be mathematically defined is producible. Machines with five axes of motion are capable of tracing contours and curves that would be difficult if not impossible to produce on conventional equipment. This capability has led to a new range of design possibilities. For example, parts that were formerly fabricated with a great deal of difficulty from simpler subassemblies are now made in one piece with greater rigidity and less weight. The bulkhead of the McDonnell Douglas F-15 begins as a 1300-pound forging and, after three months of machining, is carved into a 115-pound part. Moreover, once the program is correct, successive units are consistently produced to the same tolerances.

Finally, the flexibility of numerical control makes possible far more variety in what is produced and requires less lead time in producing it. This has far-reaching significance in the marketplace. It means the design of a product is no longer limited by the inflexibility of the technology that produces it. Moreover, a part or an assembly can easily be altered to meet the needs of different users or adapted to rapidly changing market conditions. There is already a burgeoning clamor for more customized products stemming from changing consumer tastes, material availability, energy requirements, federal regulations, and a host of other factors. Not only does NC facilitate meeting this demand but the increased availability of these products further spurs the desire for them. This in turn leads to a heavier dependence on NC.

In some cases, the key advantage is the ability to move more rapidly from the conception of a product to its production. For example, if an engineer discovers a design flaw midway through the machining of 1000 prototype automobile engines, the necessary corrections can be made without the need for major changes in jigs or fixtures or in the machine tool itself. As a result, engineering changes can be more quickly and efficiently transferred to the shop floor, thus speeding up product development.

The catch to all these benefits is that they can quickly become illusory if there is faulty managerial planning or if there are conflicts in the system. NC is a capital-intensive technology that often requires a high overhead of part programmers and other support personnel. So when things go bad, they can really go bad. In a rather startling admission at an earlier point in NC's development, John H. Greening, the president of the Manufacturing Systems Division of Kearney and Trecker, a large builder of NC machines, warned:

> It is estimated that 50% of numerical control installations are not successful because they are either misapplied or improperly supported. This failure rate parallels similar experience with computers, but these rarely show up in magazine articles and never in sales material ballyhoo.[3]

Further problems arise from the engineering fascination with new processes in general and with computer technology in particular. The danger is that engineers, mesmerized by high technology, veer off toward complex systems as a challenge rather than holding to simpler more effective approaches. The result can be delicate systems unnecessarily prone to failure. This tendency has been compounded by heavy military involvement and cost-plus federal financing in the design and development of numerical control. A 1981 Air Force report was scathing in its criticism of this tendency in weapons systems but it might as well have been talking about manufacturing technology. The study maintained that the Air Force had "evolved a self-reinforcing—yet scientifically unsupportable—faith in the military usefulness of ever-increasing technology," and went on to call this tendency "a form of organizational cancer."[4]

The desire to apply the most "advanced" technology, regardless of its applicability, is a difficult motivation to quantify, but clearly affects the adoption of new manufacturing systems. In the case of numerical control, its importance is further masked by the murkiness that surrounds the justification process for new machine tools. In spite of the fact that computerization is supposed to rationalize the nature of machining, a considerable amount of "feel" remains in justifying the purchase of a new system. This lack of hard data can of course lead to a pioneering use of new technology, but it can even more easily result in some real boondoggles. A study of NC technology in a number of U.S. manufacturing firms done by the Center for Policy Alternatives at the Massachusetts Institute of Technology comments on the paucity of data:

> One factor recurred at many of the firms: lack of hard financial data and existence of few post-audits. Although not specifically asked, it became obvious to the interviewers that the data was either not available or that the management level interviewed did not have it. In either event, equipment justification did not appear to be performed using well-developed cost projections backed up by historical data.[5]

Initially, the proponents of NC hoped that it would quickly transform all of metalworking. This didn't happen, but NC's influence is nevertheless extremely significant. In many respects, the slow adoption was due to a variety of factors: unfamiliarity with the technology, its initial expense, and the essential conservatism of machine tool buyers. Whatever the cause, a frequently cited statistic is that only 5 percent of the machine tool base is computer controlled. This statistic, however, understates the true importance of NC today for two reasons. First, the machine tool base is artificially inflated by the practice of not junking machine tools after their useful life is over and retaining them, say, to produce spare parts. If we look at machine tools less than ten years old, 7.4 percent are numerically controlled.[6] Second, each NC machine is far more productive than the conventional machines it replaces. As a result, the relatively heavy purchase of NC machines between 1973 and 1978 caused the real output of all metal-cutting machines to increase while the actual number of machines dropped.[7] A more accurate measure of numerical control's influence is its share by value of the machine tools produced. In 1979, this amounted to 31 percent of the metal-cutting machines shipped with a value of over $2,500.[8]

In some areas, NC is already the backbone of production. In aerospace, for example, 23 percent of the metalcutting machines purchased in the last ten years are NC.[9] The actual number of NC machines in use has been doubling about every five years and was 103,000 units in 1983. Most companies are planning on becoming even more heavily committed to this type of production. Caterpillar Tractor, for example, has announced that 75 percent of the new machines the company buys will be numerically controlled.[10] More significantly, two thirds of the machine tools in the admittedly somewhat inflated U.S. machine tool base are more than ten years old and one third are more than twenty years old.[11] As these machines are replaced, the diffusion of NC will no doubt increase. As the editor of *American Machinist* put it, "The day may be coming when any machine without its own internal computerized NC system will be as unthinkable as a machine without its own electric motor," and that is pretty unthinkable.[12]

NUMERICAL CONTROL ON THE SHOP FLOOR

The early expectations of numerical control's ability to remove skill from the shop floor were quite high. In a manual entitled *Beginner's Course in Numerical Control*, published in the 1960s by Cincinnati Milacron, the largest machine tool builder in the United States, the following account is given:

> So far as "operator manipulation of the machine" is concerned, it is obvious that it is greatly reduced with numerical control. The function of the numerically controlled input is to manipulate the machine from the standpoint of feeds, speeds and other auxiliary functions. . . . The number of operator variables and judgement factors are greatly reduced by numerical control.[13]

Further, management obtains the long-elusive goal of communicating directly with the machine tool. The manual continues:

> No longer are the planning functions transmitted to supervisors, routers, operators and inspectors on the operational floor in terms of descriptions, words and other English language communications. This information is now carried directly to the machine tool in the form of numbers. . . . There is almost no area left for misinterpretation or individual interpretation.[14]

What are some of the organizational consequences of these lowered skill requirements? It becomes possible to run production with fewer workers who have broad manufacturing knowledge. A skilled worker produces the first part, and a less-skilled worker makes the rest. This is the practice in the Schwartz Machine Company, a small prototype firm that is located on the outskirts of Detroit. According to Robert Schwartz, the president of the firm:

> By employing numerical control we can take a skilled man and use him as our program developer and setup man until we have the basic tape prepared and then put a lesser experienced operator on the machine for the multiple production of prototype parts. In this way we can

take an unskilled person and rapidly train him to become productive.[15]

In this regard, NC doesn't accomplish anything remarkably different than what was done to an earlier generation of production machinists who saw their skills downgraded by steel jigs and fixtures on the machine and specialized set-up men in the plant. The fact that skill is now embodied in a computer program, however, makes the application of numerical control far more flexible, and the simplification of machine operation itself makes the removal of skill more complete.

In some shops, numerical control and conventional operators are kept as far apart as possible. Recognizing important differences in job content and work attitudes, management does not want the NC operators following the lead of the conventional machinists. In one prototype shop I visited, the first job of a short production run is often made on conventional equipment to gain experience; later parts are transferred to numerical control. The NC operators are generally aware of the time it takes to produce the part on the conventional machine and see little reason to produce it faster on numerical control, machine capability or no machine capability. One supervisor complained:

> Our problems are multiplied by conventional and tape in the same building. You could push a lot harder if the tape was alone. If the operator knows it took 4 hours on a conventional machine, he figures "why should I give it to the company in 45 minutes."[16]

The NC operators in this plant also absorb the work attitudes and seek to emulate the work practices of the more highly skilled conventional operators. On the high-precision jig bore, for example, the machinist charts the job before he begins work. He generally gets a cup of coffee, sits down, spreads out the blueprints, and begins planning the work. Seeing this, the NC operator might conclude that this is an excellent way to spend the first few hours of the day. Unfortunately for the NC operator, the charting is already embodied in the part program. The same supervisor complained once again:

> A good conventional operator is a prima donna. You know that and you accept that. Back in the NC section, you are looking for very different qualities. The two don't mix.

The decrease in the need for skill can lead to an increase in the pace of the job. A frequent managerial demand is to have one worker operate two or more NC machines. Since 1960, McDonnell Douglas and the International Association of Machinists have conducted a running battle over this issue in Building 27 of the company's sprawling St. Louis plant.[17] The struggle, however, reached a new level of intensity in 1977 when the company began demanding that a single machinist run two mammoth multi-spindle milling machines. The two machine types involved, called Cincinnatis and gantrys, are truly immense. The machine tables stand three or four feet off the ground, are twelve to fourteen feet wide, and are up to forty feet long. Moreover, each machine carries four to six spindles, allowing the multiple production of complex aircraft parts. The machines were originally installed in 1969; in 1976, they were equipped with adaptive control, an automatic sensing device that adjusts the cutter speed, to further minimize the operator's role.

The company began its campaign to have one worker operate more than one of these behemoths slowly. In the summer of 1977, machinists were asked to operate a second machine for short, intermittent periods, say, while another operator went to lunch or on break. After this proved successful, longer and more complex assignments were made. This triggered a union grievance, which was bitterly contested all the way to an arbitration hearing.

The hearing on October 24, 1978, in St. Louis underscored a use of numerical control that diminished the satisfaction machinists got from their work and increased the hazards. The union complained that the distances between machines, sometimes over twenty feet, were cluttered with work materials and hoses; this made moving between them difficult during an emergency. Further, it was treacherous climbing up onto the large machine tables, which are covered with oil, coolant, and metal cuttings, to change spindles or take care of other problems. Two machines meant that an operator would be scaling one machine while watching the cut on the other and twice as often. Under normal operating circumstances, it was always possible for a cutter to fly off or for a part to break loose but under these conditions some of the machinists were more fearful of serious injury or even getting killed. The added responsibility also created significant mental strain. A machinist might be responsible for watching six expensive aircraft parts being rapidly machined simultaneously. The workers complained of feeling tense and upset, and the safety problems exacerbated the stress. The union also made a compelling case concerning the product quality. If one worker was operating two machines, there was much less chance of catching any errors and correcting them before a considerable amount of scrap was run. An operator, for example, might be in the middle of a delicate cut that requires close attention on one machine when all of a sudden a strange noise starts coming from the cutters on the other machine. Under these circumstances, there may be no good way to assess which job requires attention most urgently and little ability to take corrective action on both machines simultaneously. Finally, one unstated though very real concern was the fear of job loss should the company win the right to eliminate half the machine operators.

The company was sanguine about the safety problems, maintaining that its safety record was better than the rest of the industry. It pointed out that the control consoles for the machines were side by side and special shoes designed to be used on slippery surfaces had been provided for the operators. Should a catastrophic failure occur, the adaptive control system would automatically shut down the machine. Management argued further that any manufacturing process is dangerous, and an operator could only be injured on one machine at a time in any case. A supervisor who had been a machinist testified that there was so little to do on the NC machines that it was difficult even to stay awake. Under close questioning from his former colleagues, however, he admitted that this had not been his position when he had actually been running these machines. At the time, in fact, he had complained quite vocally about these hazards.

The arbitrator upheld the company's authority to assign one machinist to two machines, largely on the ground of managerial prerogative and some past precedents. But although the company clearly won the battle, it may have lost the war. The resentment of the NC operators at the decision makes it very difficult to regularly assign two machines to one worker. After all, a disgruntled machinist could always take a little extra time shutting off a machine during a crisis and, with machines valued in excess of $1 million and workpieces worth in the neighborhood of $250,000, that would be quite costly.

Although in theory numerical control can be used to downgrade the worker's control over the machining process, in practice there are significant roadblocks to full automation. These technical limits influence the amount of control the machinist retains and, hence, his leverage on the shop floor. To understand the nature and extent of these constraints, return to the theme of skill. In a somewhat arbitrary way, a machinist's skills can be separated into three interrelated components: conceptual skills such as planning the job, motor manipulative skills such as turning cranks and flipping levers, and feedback skills such as monitoring and controlling the actual cutting of the metal. NC is designed to bypass all three but, as will be seen, in practice fully eliminates only the motor skills in most applications. It presses down on the conceptual and feedback skills, but the door is left open for varying levels of worker input.

Consider first the conceptual skills. In theory, the programmer does the planning and the machinist executes the job. In prototype and low-volume production, however, following the theory fully would be equivalent to sabotaging production. For one thing, the part program rarely works the first time. On a particularly complex piece, the programmer might spend weeks watching trial parts cut and then shuttling between the plant floor and the engineering office to make needed corrections in the program. Some of the errors are obvious, particularly those resulting in a smashed cutter or wrecked machine, but some problems are not so obvious and the machinist is an important source of ideas and information. At the machine every day watching raw stock shaped into finished parts, the worker's cumulative experience, insight, and "feel" are invaluable in suggesting, say, a better cutter or seeing a faster cutter path or explaining why the metal is being gouged when a good finish is essential. David A. Price, the manager of technical services at Manufacturing Data Systems, Inc., a firm specializing in computer part programming, underscores the value of this input:

> A good operator can, if he wishes, make good parts from a less than perfect tape until the programmer can correct it. He can work with the programmer by giving advice on machining methods, cutting tools, and conditions on machines requiring special knowledge.[18]

As a result, some programmers seek to consult and collaborate with machinists regardless of the formal procedures involved. At issue, however, is a question of power: Who controls the machine tool? The programmer and not the machinist is usually empowered to make the final decision. In fact, management at some operations seeks to use the programmer as a filter to let the good ideas of the machinist through while limiting the operator's ability to disrupt production by retaining the final say. The consequence of this approach is a collaboration that is hardly congenial and one that can throttle productivity. Further complicating matters is a perceived class difference between programmers and machinists. When the programmer comes down from the engineering office to the shop floor, there is often a confrontation between feelings of white-collar superiority and blue-collar resentment. Management at times exacerbates the hostility by stressing the difference. Richard Palmer, a young NC operator, describes his feelings:

> Really, what's happened is that you're not running the machine alone—there are three or four people running it—the engineer, the programmer, the guy who made the fixture, the operator. . . . One thing that happens is that it's too hard to communicate with the other people involved in the process. They don't want to hear it. They've got all the training, all the degrees. They just don't want to hear from you about anything that's gone wrong. It's got to be all your fault. They sure won't admit it if *they've* made a mistake. . . . When I do find a way to improve some operation, if I can do it without anyone seeing, I don't tell anyone. For one thing, no one ever asks me.[19]

Ken Fairn, an NC machinist at Rolls-Royce in Britain, also comments on the potential for hostility between programmers and machinists. "If you've been a skilled man for years and years you could tell them what tools to use but they don't ask and then you get so you don't want to offer."

In medium-batch production, where hundreds of the same part are made at the same time, the operator is particularly vital in removing bugs from the system at the beginning of a production run. In a study of a new automated department at Caterpillar Tractor's mammoth plant in East Peoria, Illinois, Clint Stanovsky, a former researcher for the Center for Policy Alternatives at MIT, describes the expanded role of NC operators when both the production run and the machines are new:

> In fact, during the start-up of the new systems, the programming bureaucracy at Caterpillar has not been able to keep pace with the need to change programs to compensate for variations in castings, in anticipated tool behavior and wear, and to reflect engineering changes in workpieces. Factory managers agree that if operators could not program their machines on site it would be impossible to meet production goals while phasing in the new equipment.[20]

Once the system is operating properly, machinists with too much experience and skill may be viewed as a handicap. In the Massey Ferguson plant in Detroit, managers echo the engineers in the early days of the Ford Motor Company who preferred machine operators with no experience and therefore no preconceived idea of what to do. In this case, the numerically controlled lathes are physically very similar to the cam-operated automatic lathes they replace. Doug Keno elaborates:

> With NC, the less they know about manual machines the better. The whole concept is completely different. Very little feel is left for the operator. You tell him what tools and what position. There is very little left for his judgment.

Of course, the less knowledge the worker has, the more control management retains. If the tool engineer wants the tool to last for ten minutes, the machine is programmed to stop after ten minutes and automatically signal the need for the tool to change, rather than relying on the machinist's judgment. An experienced operator is a handicap because he might resent this and try to intervene anyway. According to Keno,

> an older operator may feel degraded on the machine. His pride may be affected. I took one operator with 25 years on the turret lathes. In a week he was a nervous wreck because the machine was doing everything he had done before. Someone like this might push the wrong button accidentally or on purpose.

Blocking this intervention, however, can be costly. In a number of perceptive case studies on the introduction of new technology in manufacturing, Leslie Nulty, a former staff assistant to the president of the International Association of Machinists, highlights why this approach can be counterproductive. One case Nulty looked into was a small family-owned firm with fewer than 100 people on the shop floor. Organized by the IAM in 1946, there has never been a strike at the plant. At the time of the study, the firm had four computerized machine tools: a jig mill, a shaft lathe, and two chuckers. The machines were brought in to raise production as quickly as possible during a period of rapid business growth, increasing competition, and rapidly rising backlog. Once installed, the machines were pushed to their limits, with little regard for worker suggestions on machine operation or other factors such as preventive maintenance. According to Nulty,

> worker suggestions that better maintenance might lead to better production have been shrugged off by management despite the fact these are $350,000 machines pur-

> chased with loans at 18–20 percent interest rates. The members have an explanation, in their words: "You know—workers are the lowest form of animal life."[21]

The consequence was premature failure on the machinery and costly downtime. While workers were ignored before the fact, "after two or three years, management learned that the maximum is not necessarily the optimum and its engineers reduced 'standards'—the designated machine speeds given the workers for each type of material and part."[22] Surprisingly, preventive maintenance is still largely ignored, which not only affects production but increases stress on the job. Poorly maintained machinery is more likely to cause an accident, say a part flying off the machine, and makes it more difficult to produce within tolerances.

Success in removing conceptual skills from the shop floor—in further severing planning from execution—creates some new problems. The link between the machinist and the design process is broken. The dialogue between the engineer and the person closest to the cutting and fabricating of metal is not mediated by the part programmer. Not only does this restrict the upward flow of design ideas, it may limit the engineers' awareness of the problems involved in producing a certain design. The result can be the exploration of fewer design alternatives.

Now turn to the feedback skills. The NC program is in many ways similar to guiding a car by remote control. Imagine selecting a route with the aid of a map and determining the vehicle speed based on suggested speed limits for each part of the journey. Serious problems can arise if there is no way to monitor road conditions while the trip is in progress. If it is unexpectedly raining or if there is an unanticipated crater in the pavement, for example, the car will nonetheless plow ahead at its full programmed speed. If a bridge happens to be washed out, well so much for the trip and probably the car. In prototype work in particular, the variety of different metals and cutting operations creates constant uncertainty. According to a programmer in a General Motors' die shop,

> the engineer or programmer may think that he is responsible but there's no way that he can control what happens when the metal is cut. I don't know a programmer that could select a feed or speed to cut aluminum bronze without a lot of luck. The operator has much better control than we'd ever have.

In longer run production, it is possible to optimize cutting conditions more accurately, but even under these circumstances, the unexpected can happen. As a result, NC machines are equipped with a feed rate override switch so that the machinist can adjust the programmed feed of the machine to the actual cutting conditions. If there is a hard spot in a casting, for example, the machinist can slow the feed down until the cutter passes the problem area. This technical necessity creates some real managerial concern since it reopens the loop of manufacturing control. *Modern Machine Shop* magazine lamented that "some operators will slow down the feed and lengthen the program time, as they jokingly refer to it as a 'job security' switch," meaning that work can be stretched out to secure employment in slack times.[23] At one plant I visited, the shop manager complained about a machinist who consistently ran his machine at 75 percent of the programmed speed. When the program itself was reduced to 75 percent of its former speed, the operator simply ran it at 75 percent of the new speed. At another shop, one operator ran the machine at 60 percent of the programmed value, removing the dial and recalibrating it at 100 percent for the benefit of anyone who happened to wander by. The shortfall of parts was apparent at the end of the day, but the managers had a problem pinpointing the cause. According to Clarence

Beyer, a supervisor at a medium-size machine shop near Detroit, the override switch becomes particularly troublesome if "the operator is upset because he didn't get a raise."

Some of the latest NC machines have a provision that makes the override switch inoperable. Management, however, is generally reluctant to use it because if a hard spot in fact does appear, it could shatter the cutter, scrap the part, and damage the machine tool. Moreover, using the "lock out" switch eliminates any possibility of improving a program in operation. In the Nulty case study, the results might have been worse had the workers in practice not used "their knowledge and skill to reprogram machine feeds and speeds" according to their standards rather than those of the engineers. Further, if a machinist is running behind on an important job, the override dial can be used to speed the program up.

These potential benefits notwithstanding, the drive is to remove even this level of human input. One way to do this is called adaptive control, a device that automatically senses cutting conditions and adjusts the machine accordingly, eliminating the need for an override dial. Since the cutting process is so imperfectly understood, this is a tall order. In fact, two decades of research on adaptive control have produced much controversy and few effective applications. Units that work well in the laboratory seem to go haywire once they get on the shop floor. One major aerospace company was forced to remove adaptive control from its NC machines because of repeated technical problems. *Iron Age* summarized the situation: "The ongoing drive to improve productivity in metalworking worldwide has yet to uncover an innovative technology that has promised more—and delivered less—than adaptive control."[24]

An alternative possibility would be to develop adaptive control as an aid to the machinist. The system could shut off the machine in case of a catastrophe, for example, but under normal conditions provide a range of choices for the machinist to fine tune the program. Or, the technology can be viewed as a total replacement for the worker's input. Given the enormous technical difficulties involved, the former goal would seem most sensible. Sensibility, however, is not the determining factor. The target, as in other technological refinements, is whatever resistance workers have shown to previous versions of a system. As some leading managers in the aerospace industry put it, "in most recent systems, the feedrate override is locked from operator control during adaptive control cuts so human intervention is avoided."[25] Douglas Schenck, a programming manager at McDonnell Douglas, elaborates on this theme:

> Adaptive control takes the override dial away from the operator. It stabilizes the run time. Before AC, the first shift might run ten parts, the second shift eight to ten, and the third shift two. It was difficult to forecast. Now the operator can't slow the machine down. Unfortunately, our adaptive control is strictly a slowing down mechanism. If we put the feed rate too low, that's our problem.

The organizational implications of adaptive control are delineated by Richard Johnstone, vice president and general manager of Kearney and Trecker's Electronic Products Division:

> These sophisticated features . . . force the user to define what he intends to do. Part making is being planned ahead of the event not corrected after the event. And we're bringing the responsibility for heads-up manufacturing planning off the shop floor and up into the production engineering office where it belongs.[26]

But even without an override switch, the machinist retains some ability to pace the job. In one shop, for example, when an NC operator prefers not to load another part near quitting time, he instructs the cutter to trace the path just cut, in effect "cutting air" with all the precision of computer control. (Frederick W. Taylor might have called this "automatic soldiering.") Moreover, although the cutting process is supposedly fully automatic, manual intervention is nonetheless required at a number of key points, say checking the part's dimensions with a precision micrometer. In addition, the machinist is frequently responsible for bringing the tooling to the machine and doing the required setup. The difference between the time theoretically estimated to make the part and the time actually spent on the job is often covered by a general category called "unavoidable delays." The length and frequency of these delays not only indicate the number of technical problems encountered but also reflect the level of disputes between workers and managers. If machinists are being pressed, unavoidable delays may increase. One NC operator describes efforts to increase his output on a new machine:

> The shaft I was making weighed about sixty pounds. Everything was all right until they decided to send time study down on the shop floor. They send a kid 25 or 26 years old who says I should be doing 104 pieces for 8 hours. I tell him "hey man, I'm not Superman," because I was only doing 70. I screamed and yelled and told the foreman that this kid is dreaming. So they send another time study guy down who has more experience. He knows about "unavoidable delays," et cetera and he comes up with 91 pieces for 7 hours. I'm still doing 70 and I'm not going to do one more. I figure 70 is pretty good for the job since it's a problem job to load and you get pretty tired out by the end of the day.
>
> Since it's a new machine, the plant manager comes down every day and asks "how's the machine running"

> and people are always drifting in and out asking you questions. The foreman doesn't say much. I cover myself on the job ticket. I have one and a half hours down with unavoidable delays. The foreman knows that if he presses me I can always start turning out 50 pieces, and not all good ones. The unavoidable delays have nowhere to go but up.

There are two further roadblocks to extending managerial authority via numerical control on the shop floor. First, NC does not directly affect what happens when a workpiece is not on the machine. In most shops, parts spend a considerable amount of time not being worked on, either being transferred from one machine to another or lying around waiting for available manpower or machines. (This shortcoming becomes a potent incentive to automate far more operations, of course.) Second, NC leaves untouched important aspects of metalworking trades such as diemaking. Only 40 percent of die construction, for example, involves machining and not all of this by any means is NC machining. There are, of course, significant indirect effects. The greater precision of numerical control minimizes subsequent hand fitting, grinding, scraping, and polishing and therefore diminishes the need for these skills. Control over these remaining operations, however, remains with the diemaker. Further, the complexity of diemaking has resulted in a slow diffusion of NC. In fifteen major General Motors die plants, for example, only 15 out of 2,000 machine tools are numerically controlled and half of these are in one plant.[27]

Nonetheless, one small midwestern machine shop has used NC to subdivide the production of simple dies into such narrow categories that a leading trade journal has referred to the process as "diemaking by the numbers." Each NC operator is responsible for certain machine operations that are numerically coded. The plant superintendent subdivides and numbers the needed diemaking operation and a production coordinator links people and machine work "without necessarily knowing what a die is."[28]

This small example underscores the fact that despite its shortcomings, numerical control lays the basis for a further fragmentation and intensification of skilled machining. Human input remains important, but the nature of that input becomes more and more reactive. The skilled worker no longer has a great deal of autonomy and judgment about initiating the job, although the ability to improve the program or foul up the operation remains. Even as powerful a weapon as the override dial only allows the worker to select a certain percentage of the management-determined time to do the job. And, as the technology develops in directions such as adaptive control, the machinist is caught between ever tightening managerial constraints.

One far-reaching extension of NC technology is the unmanned machining center. This technology combines an NC machine, an automatic cutting tool changer, and a connected series of pallets to hold parts. It is not quite unmanned. On the first shift, a machinist loads the parts to be machined, inserts the cutting tools, makes any necessary adjustments, and watches the operation for eight hours or so. On the next two shifts, the machine theoretically operates unattended. The computer program triggers the necessary cutter changes and, when one part is done, the next pallet automatically slides into place. Although unable to machine the most intricate of parts, this type of system represents an important technical breakthrough and is being widely promoted. In the advertisements for the Cross and Trecker version, a caveman is shown chipping a rock under the caption FIRST BREAKTHROUGH . . . MAN, THE TOOLMAKER. Underneath this is a photograph of the latest unmanned machining center with the caption LATEST BREAKTHROUGH . . . THE TOOL THAT WORKS WITHOUT THE MAN.

The necessity for varying levels of human input with NC, even in the unmanned machining center, creates some complex pressures that influence the way the technology is used. Unable to lock the machinist out entirely, management is generally fearful of fully involving the worker. One central question is how skilled a machinist on the job must be. In the early days of numerical control, the need for less planning and fewer motor skills led to the widespread expectation that anyone who could push a button could operate an NC machine. Subsequent experience, however, has indicated that more is involved than operating an automatic elevator, although obviously fewer skills than operating a conventional machine tool are needed.

There is now widespread managerial agreement that the operator must be responsible, alert, and motivated. For one thing, there is often a considerable investment in the machine tool and the part. As Burnell A. Gustafson, a Litton Industries machine tool executive, put it:

> If you have a $500,000 machining center, you don't want some clown pressing the wrong button, even if the machine has all the fail-safe devices in the world built into it. I don't see us getting to the level where the untrained worker can step up and run the equipment by pushing a button.[29]

The quality of central concern to management is the "attitude" of the machinist. An early study of numerical control in the Southern California aerospace industry stressed motivation as being critical.

> Of all the characteristics desired, motivation—interest in advancement—was the most important quality an individual could possess. Education, training, and experience hurdles could be overcome if a person were

> sufficiently motivated. On the other hand, lacking proper motivation, these requirements could constitute insuperable obstacles. Other attributes of the "ideal" employee related to his stability and reliability. Married men with families were often preferred as reflecting such attributes.[30]

Clint Stanovsky's study of automation at Caterpillar Tractor also emphasized worker attitude as being of central importance to the company.

> "Good attitude toward the company," interpreted to mean the will and ability to act responsibly on behalf of the firm, was the primary criterion used by managers who selected operators for the new jobs.[31]

Since responsibility and alertness are qualities that are generally found in skilled workers, some managers have opted toward having skilled machinists run NC machine tools. From a management point of view, the major costs on a million-dollar machining center are associated with the capital investment and not the hourly labor cost to operate it. The extra money spent for a more skilled worker who is intimately familiar with the metalcutting process is viewed as a form of insurance. In other cases, unions have successfully pressed to retain skilled workers on NC equipment. The experience of some managers, however, has led them to believe that the insurance is not worth the premium. For example, the assistant director of employee relations at a West Coast aerospace company maintained:

> Basically, when we started in using NC equipment we were experimenting. We started with machinists that were on a similar level with the other machinists. We found that it didn't require a master machinist's skill to run

> these machines but a much lower level man because he was only required to monitor the machine in case of a problem. Because of this we have brought in people with no background whatsoever in machine shop and they have been able to run these expensive machines. In the manufacturing department there is on-the-job training, the previous or present operator teaches the new man how to run the machine and what to do in case of a problem.[32]

This strategy, however, can create some devastating long-term consequences. The less skill required on the job, for example, the less skill available when it is vital for production. A programming manager in the McDonnell Douglas, St. Louis complex describes a catch-22 situation:

> NC was developed to an extent to substitute for the lack of skilled operators that was foreseen. As a result, very few machine operators can really assist you in evaluating a part program because they don't have the skill. People who have the skills want to do something more meaningful.

For some companies, securing the "right" operators is a matter of considerable strategic importance. In unionized plants, however, this goal may conflict with seniority agreements that govern promotions and transfers. Under these agreements, if a number of workers meet the qualifications set for a new position, the worker who has been on the job the longest is entitled to it. If there is a layoff or other employment disruption, workers with the most seniority are laid off last, and generally have the right to "bump" workers with less seniority on lower job classifications. Years of service not attitude is the governing factor. This represents a serious obstacle to managers concerned with recruiting young, highly motivated, and management-oriented workers. Caterpillar Tractor was not content to accept these limits in staffing a new, highly automated department in an existing building in its massive East Peoria complex. According to Stanovsky's study, some managers have sought to undermine the seniority provisions of their collective bargaining agreement in order "to recruit operators upon whom they can depend to exercise greater responsibility in the interests of the firm than would be expected of operators chosen according to established wage and allocation systems."[33]

To this end, the managers of the new automated area have manipulated the job classification system and have developed informal promotional arrangements to "isolate" the new area from other workers in the plant. The plant manager, for example, fought for and won a higher classification for an automated production line than the corporate job analysis and labor relations staff thought was appropriate. A higher labor grade means fewer workers can "bump" into the classification during a layoff and allows more discretion in hiring workers initially.

In itself, the formal job classification system might not be enough to "isolate" the department. But, the mystery and fear that new technology engenders lay the basis for weeding out many prospective applicants.

> The Factory Manager believes that most of the older operators are discouraged by the unfamiliar tooling, tolerances and controls that they encounter in the new department. Supervisors have had more success with younger employees and recruit them more vigorously.[34]

The initial fears and uncertainties of new technology could be allayed, but since the older, more experienced workers are not particularly wanted for the job, the mysteries of the technology provide an effective barrier. Recruiting younger workers more vigorously undoubtedly means that countless informal obstacles are erected for the older workers. For example, knowing when and where to apply for a promotional job and what qualifications are desired can be helpful but may not be forthcoming.

Some managers, however, complain that the new pay grades are not high enough to insure the uninterrupted supply of the type of workers they want. Stanovsky disagrees, pointing out that there may be a hidden reward to only moderately higher pay grades: The system attracts highly motivated employees without creating deep resentments among those that are passed over.

> Diminished rewards for promotion to new labor grades are also diminished losses for operators who are not selected for the new jobs. Such low incentives for promotion allow managers freedom to pick and choose without arousing disappointment and grievance activity by operators who are passed over in the selection process in favor of employees with less seniority. Employees who wish to operate the CNC systems are likely to have either non-financial or long term career motives.[35]

Management may, in fact, be using the introduction of new technology as a convenient vehicle to secure tighter control over a section of the work force. The Caterpillar study maintained that the workers in the automated department had no more responsibility and in fact required less skill than the machinists in the two conventional departments studied. Rather than an "elite" work force being technologically mandated, it was in fact socially desired by Caterpillar management.

> Installation of automated production equipment facilitated changes in work structure by obscuring the nature

> of the change. The new machines and their requirements are largely unknown to local union leaders, and certainly require new knowledge to operate. This knowledge is available first only to managers, who can choose how to train operators for the new jobs. In so doing managers have had an opportunity to define the tasks and qualifications of those who perform the new work, and have taken the opportunity to tighten their control over the allocation process.[36]

Plant management appears to have been successful in achieving its immediate goals. The workers in the automated department are younger and have on average much lower seniority than their counterparts in the two conventional departments. While more of the machine operators in the larger plant in which the automated department is located have had a formal opportunity to move into the new department, few have taken advantage of it. According to the study:

> The personnel of Department 163 reflect the desire of the factory managers for an elite group of highly motivated employees willing and able to learn the new machining technology. Their extensive participation in company training programs suggest that these operators are potentially well integrated into the firm and probably safe recipients of the trust their managers place in them.[37]

In the long run, however, the attitudes and performance of the workers in this department may change in ways undesired by Caterpillar management. After all, General Motors had recruited a young and highly motivated work force to staff its Lordstown assembly plant in the early 1970s.

PROGRAMMING AND COMPUTER NUMERICAL CONTROL

Numerical control splits what a machinist does on a conventional machine into two parts. Having discussed the operation of the machine on the shop floor, we now turn to programming, the development of the instructions that guide the machine tool. The design or selection of programming methods is an involved process that depends on quite a number of technical factors, from the complexity of the workpiece to the resources of the workplace. A central goal, *Modern Machine Shop* reminds us, is to "handle the workload of a plant at the lowest possible overall programming and machining cost."[38] Technical and economic considerations alone, however, don't tell the whole story. The selection also depends on some critical social choices. In effect, decisions about how to program the machine tool are choices about how to organize the workplace. They influence how much skill and control the machinist retains on the shop floor and where in the management hierarchy key production decisions are made. *American Machinist* magazine tells us that "in most NC installations, the programming function is removed from the shop floor and from the machine operator."[39] Whether this is technically mandated or a reflection of social purposes is frequently obscured because issues of managerial authority and technical necessity are intertwined. In this discussion I will try to unravel these potent themes by first looking at the process of programming in more detail, then by examining who does programming and why, and finally by exploring some new technical developments in numerical control and their implications for the organization of work.

A part program defines the final shape of the part and spells out the process that will be used to get there. It is a symbolic statement of the entire machining operation, detailing the sequence of cuts, the cutters to be used, the feeds and speeds, the path of the cutting tool, and anything else that is supposed to be taking place. The complexity of the program and the sophistication of the programming method depend principally on the geometry of the part. A simple bolt, for example, may need only a few machine instructions while the bulkhead for a jet aircraft could require a very high-powered programming system and hundreds of thousands of instructions. The more involved the method, however, the more difficult it is to use and the more likely that its use will be separated from the shop floor. In particular, a sophisticated method used on simple parts can needlessly restrict machinists from programming. Analyzing all the factors that lead to the selection of a given method is much beyond this discussion but it will nonetheless look at some of the broad approaches that are used.

One important choice is between manual and computer-assisted methods. Manual programming is generally used for less-complicated parts and when less programming has to be done. Basically, a programmer describes the overall geometry of a part and what the machine is supposed to do in a structured form that is readable by the machine. When the computer is introduced, the programmer works with a "higher level language," a strange-looking amalgam of English words and symbols that makes possible vast new machining capabilities and requires far fewer calculations. Although there are fundamental technical differences between manual and computerized methods, the introduction of the computer does not by itself mean that programming is any more or less removed from the shop floor.

There are literally dozens of computer-based programming languages. Historically, the most important is APT (Automatic Programmed Tool), the forerunner of the other languages and a part programming approach of enormous capability and range. Originally designed to program the most elaborate of aircraft parts and developed with considerable

Air Force funding, APT is now available in a variety of different forms depending on the application. In its full-blown version it is powerful enough to program virtually any surface that is mathematically definable. The flip side of this versatility is its cumbersome nature. It is difficult to learn, difficult to use, and requires a considerable amount of expensive computer power to operate. In many applications it is similar to using an M-1 tank to drive to work. It will get you there but at quite a cost. As a result, APT is in limited use, primarily in large shops that have the technical and financial resources to implement it. A new generation of APT-based languages, however, has been developed to fill the technical void and to meet the needs of small- and medium-sized shops. While far simpler than APT, they are still highly structured approaches.

A more flexible language is COMPACT II, the brainchild of Manufacturing Data Services Incorporated, an Ann Arbor, Michigan, company that is now a subsidiary of Schlumberger, the French conglomerate. MDSI has parlayed COMPACT II into the cornerstone of a $60 million-a-year business specializing in software for numerical control applications. MDSI claimed its language was used to program almost half of the 35,000 or so computer-programmed NC machines in the United States in 1980. Like some of the APT systems, smaller shops are able to use ordinary telephone lines to tie into two or three central computers around the country to process their programs in a time-sharing arrangement. Although COMPACT II is still a highly structured language that requires some time to master, it is nonetheless quite accessible to machinists, opening up the potential for shop floor uses.

Other languages go much further in trading capacity for increased ease of use. PROMPT, for example, asks the programmer "what next" questions on a TV-like screen. When the programmer answers one question, the system asks for more information until the program is complete. PROMPT's backers maintain that the language is so simple even managers will know what's going on.

> Its simplicity and speed of learning will be of obvious importance to the harried executive who is often held hostage by his inability to take the time and effort required to learn the black art which traditional NC programming languages are regarded as being.[40]

An important danger for managers of conventional programming languages is that in limiting the leverage of the machinist, a dependence is created on a new skilled worker, the programmer. In particular, the owner of the small shop might find that he is, according to PROMPT's promotional literature, "highly dependent upon the knowledge and skill of a man who could not be readily replaced if something happened to him."[41] These fears recall the apprehensions of an earlier generation of managers that engineers would dominate the workplace if "scientific management" were fully adopted.

In machine shops, a key issue is who should do the programming. High on the list of qualifications is machining background, which should come as no surprise since "a programmer is a machinist without a machine tool," as one supervisor at McDonnell Douglas put it. On some jobs, the configuration of the part may be so byzantine that APT, with all its firepower, is necessary, in which case advanced engineering and mathematics may be more important than actual machining experience. But this is the exception rather than the rule. Dennis Howard, director of training for MDSI, an organization that has trained thousands of programmers, prefers to start with a machinist.

> We prefer to train a machinist because Compact II programs use machining terminologies. The program looks just like a conventional operation sheet that a machinist uses. The system is an extension of a machinist's logical processes.
>
> All he has to do is write down in words the process to machine the part or component. He doesn't have to know

> anything about a computer. All he needs is knowledge of machining and a general idea of how a typewriter keyboard is laid out.

MDSI runs classes the year around to train new customers in the fundamentals of COMPACT II, a process that takes about one week.

The value of a machining background is not limited to COMPACT II. Edward F. Schloss, a Cincinnati Milacron sales vice president, feels that the shop floor is an excellent place from which to recruit programmers.

> We've had excellent success with good lathe operators and good boring-mill hands. They don't know it, but they've been programming most of their working lives, and they know basic shop math and trigonometry. You can teach them programming rather handily. Conversely, though, it's fairly hard to make NC part-programmers out of high-powered mathematicians. The path programming is easy. But what to do with it—the feeds, speeds, etc.—that may take even more extensive training.[42]

The Moore Special Tool Company, a much smaller manufacturer of precision machine tools, recounts similar experiences. Stephen Liscinsky, the director of NC training, comments: "When I'm asked who makes the best programmers, I say that it's usually the people who have a decent knowledge of jig grinding and shop math. The important thing is knowledge of the machine."[43]

Ironically, while there is a desire to have programmers with a machining background, there is an aversion to having machinists on the shop floor involved with programming. That would conflict with managerial authority. Donald Smith, chairman of the Industrial Development Division of the University of Michigan, maintains that it would be "very undesirable to have the operator do any programming. This would take away control of the production environment." On an operating level, R. E. King, co-owner of Standard Tool and Die Company in Los Angeles, shares these sentiments. "I don't want any operator fooling around with programming. That should be done only in the engineering department."[44]

The conflict between the need for machining know-how and the desire for social control is often resolved either by making programmers out of managers who were machinists or by promoting machinists into a new position that takes them off the shop floor. This approach was very much in evidence at a one-week training program I attended to learn how to use COMPACT II, sponsored by MDSI at a location near their Ann Arbor headquarters. All twenty or so members of the class had begun their careers cutting metal but none were currently machinists. Instead, they were foremen, line managers, planners, or shop owners. When asked to characterize the programming practices of its clients in independent job shops, one MDSI executive quipped "in the small shops the owner does the programming and in the larger shops the owner's son does the programming." Moreover, in some cases, the machining background of the programmer is viewed as limiting the need for input from the shop floor. As Bob Nordhuff of Douglas Aircraft put it:

> When all the programmers have a machine shop background (as they do here), we don't need any input from the machinist. If we want to know what's happening on the machine we just send somebody to check what's happening.

In a union shop, removing programming from the shop floor generally means that the programmer is removed from the union. When NC was first introduced, this potential erosion of the bargaining unit sparked hostility and quick reactions from a number of unions. One such challenge from the UAW resulted in a landmark arbitration decision on September 11, 1961. The case was triggered when management installed a new numerically controlled drill press in the Tool Room at Fisher Body Plant 21. Given the issues of managerial prerogative involved, it is somehow fitting that Plant 21 is located in the shadows of the mammoth General Motors headquarters building on the near east side of Detroit. The case began when the corporation assigned the programming of the new machine to a production engineer rather than a toolmaker. The local union challenged this in the grievance procedure and fought it through to an umpire decision, contending that the equivalent of programming has always been done by toolmakers and "involves the highest skill of the trade." Aside from some rather narrow technical arguments related specifically to this incident, management based its case on managerial prerogative. There is something ironic about the most powerful manufacturing corporation in the world insisting that the location of drilled holes on a steel plate is a "fundamental right of Management." The umpire's summary of management's argument concerning this point is worth quoting in its entirety.

> *When the Production Engineer makes a decision with respect to sequence of operation and where the part will be drilled or machined, he is carrying out a basic Management function.* His decisions are the expression of the right of Management to maintain the efficiency of employees and to determine the methods, processes and means of manufacturing. The fact that these decisions are transmitted to the machine floor by means of a tape is of no significance. The same order could reach the floor by means of work line-up order, prints or verbal orders by supervision. The important point is that the right to make these

> decisions is a fundamental right of Management. *What the Union is asking for here is the inclusion in the Bargaining Unit of a measure of control over the methods, processes and means of manufacturing as related to the Burgmaster Machine* [emphasis added].[45]

Ruling in favor of the union, the umpire concluded:

> With respect to the machine in question, Management has simply taken the function away from the Tool Maker. If it may do so in this instance, in the interest of efficiency, it could make a similar decision as to all programming, or, indeed, all functions of the Tool Maker, or for that matter, any function previously performed by any classification. By this process the representation rights set forth in Paragraph 3 could be nullified and the bargaining unit eroded.[46]

In the twenty years following this ruling, the UAW has seen the intent if not the letter of this landmark decision circumvented. The victories won by the union have often been undermined by the corporation's skillful use of a slight change in the technology. On June 19, 1978 Irving Bluestone, the vice president in charge of the union's General Motors Department, sent a forty-six-page booklet to every General Motors' local union; in it, he reviewed the contractual language the union had won concerning numerical control and expressed concern over its repeated violations. In a cover letter to the booklet, Bluestone commented:

> A significant portion of technological change in the last two decades has involved numerical control procedures and an increased use of computers. Their introduction into any given plant has been gradual and frequently without the Union's knowledge. The Corporation has failed

> to hold advance discussions with the International Union when there is an impact on the scope of the bargaining unit as required by the Statement on Technological Progress. In many instances where the Union has become aware of the introduction of numerical control and computer technology, a substantial part of the work has been placed into the bargaining unit initially or returned to the unit when improperly assigned elsewhere.[47]

A typical example of some of the problems surrounding numerical control in General Motors' plants is found in a case that arose in the experimental machine shop at the sprawling Oldsmobile plant in Lansing, Michigan.[48] The local union complained that management was locking the computer and programming facility on the afternoon and midnight shifts so that machinists would not have access to it. The union had won some programming rights in its local agreement but the company evidently wanted to insure that these rights were carried out only in the presence of more extensive supervision on the day shift. In this way, any independence exhibited by union programmers could be contained. But for the machinists on the floor this made work more difficult. If an error were discovered on a tape in the late afternoon, for example, the job might have to be stopped until the following morning when the tape could be corrected in the computer room. Further, a machinist needed ten to fifteen years' seniority to be eligible for the day shift. This excluded from programming many of the younger workers, who were particularly anxious to learn how to operate all phases of NC technology. The practice is clearly inefficient for management but the increased control stemming from carefully supervised day-shift programming was apparently worth the price. In addition to a formal union complaint, a numerical control machinist wrote a lengthy letter to management setting forth his grievances:

> Management feels that hourly employees would destroy and sabotage programs and equipment.
>
> There are security systems that can be set up for the computer to protect itself and the programs on file. These systems still give access to the operators. I don't know which one would be best but I am sure there is one that would work for us and not against us.
>
> As for the hardware itself, it is in no more danger than any other machine in the shop. . . .
>
> This would presume a *small* amount of trust of its employees by management but I don't think this is too much to ask.

The question of who should do the programming is closely linked to where the programming should be done, on the shop floor or at a remote site such as the engineering office. The latest generation of numerical control—computer numerical control—opens some new possibilities for organizing the workplace, and as a result underscores the tension between management's need for more flexibility and its desire for increased authority. With conventional numerical control, it is impossible to produce or alter a program at the machine tool. This means that correcting or improving the program has to be done off the shop floor, an inefficient, time-consuming, and frustrating process. It would be faster and easier to make needed changes and test them out immediately. CNC lays the technical basis for doing this through the use of a minicomputer at the machine. A new program can be developed or an existing one modified with the ease of altering a cassette on a tape recorder.

CNC systems were particularly in evidence at the 1982 National Machine Tool Builders Show in Chicago. In fact, they totally dominated the control systems on display, appearing on almost every machine. At first glance this would appear to be a bonanza for those interested in returning control to the shop floor. But, as we will see, the possibilities also exist to utilize CNC as a vehicle to remove skill from the machinist's job. In this case, much depends not only on the design of the system but on the way in which it is deployed and used.

There are really two types of computer numerical control, one simple and the other more complex. The less sophisticated type is generally called manual data input (MDI) and is designed to be programmed at the machine tool. The program is entered into the machine via a keyboard resembling a sophisticated calculator. The machinist looks at a blueprint, enters the proper dimensions, and programs the part. While this is often referred to as the operator doing programming, it is no more like part programming than recording an airline reservation is like computer programming. Essentially, the computer automates much of the programming process and does it right at the machine tool.

With the more complex variety of CNC, the capability of the machine controller approaches the programming ability of a traditional NC language although with less skill involved. Some of the units are "conversational." The operator enters a command such as "bolt circle" and the machine in return asks questions such as "how many holes?" One such system manufactured by General Numeric was displayed at the 1982 machine tool show. The technical literature describes some of its attributes:

> This unique, new General Numeric System P–Model F programming computer revolutionizes NC part programming. It is the first programming system that permits operators with absolutely no knowledge of NC programming to generate a program—merely by pressing buttons.
>
> There is no need to know, or even to learn NC programming, nor is there any need for calculations. They are performed automatically by the System P–Model F.

> All the operator must learn to do is respond to the questions which appear sequentially on the CRT screen of this interactive, conversational system. The part profile is then displayed on the CRT, as the input dimensions from the blueprint are given. Symbolic figure keys take care of the profile of the part.[49]

When the System P–Model F was demonstrated at the General Numeric booth at the tool show, the company also stressed a less obvious feature. The machine control unit is set up to project the theoretical time to perform each operation so that this figure can be compared with the actual time spent on the job. In fact, this projected time is broken down into several categories such as the time the tool spends cutting metal and time spent moving between cuts. The surveillance possibilities are endless.

At the Kearney and Trecker exhibit a new CNC system called SETUP was demonstrated. SETUP stands for Shop Entered Tutorial Programming. It provides the capability for an operator to program a part while another completely unrelated part is being machined. From the machinist's point of view, however, a system such as this could easily become a means to speed up what a worker does rather than a way of preserving shop floor autonomy. Nonetheless, the programming capability is certainly there.

The amount of input from the shop floor is very much up to management and as a result the technology is used in a variety of different ways. In some cases, such as the die room at the Rouge plant, diemakers fully program simple MDI machines. In other cases, a supervisor or shop owner will program the first part and then turn the machine over to a less skilled operator for the remainder of the production run. One machine tool manufacturer calls its control unit Journeyman, perhaps for its role in potentially replacing skills. In an ad directed to the owners of small shops, the manufacturer of Journeyman describes its potential uses under the banner headline "SKILLED HELP HELP."

> This is the Tree Journeyman. It's a new kind of "programless NC" machine that solves two kinds of skilled manpower problems.
>
> If you have skilled help running manual machines, Journeyman lets you run the first part using manual data input and fast, automatic positioning. From the second part on, the "memorized program" takes over. When you've finished your run, you can transfer the program from memory to tape cassette for use on repeat jobs.[50]

Another ad by the same company makes the point even more directly with a large headline that proposes "YOU MAKE THE FIRST PART. YOUR JANITOR CAN MAKE THE SECOND ONE."

With more complex CNC systems, the machine capabilities equal or exceed those of conventional NC. As noted, the decision to program on the shop floor is based on a number of technical and social factors. Regardless of where programming is done, however, it is generally an important advantage to edit the program at the machine tool. This avoids delays and, often, considerable confusion. The machine control, of course, doesn't know if it is the machinist or the programmer who is doing the editing. Clarence Beyer, a programmer and supervisor at the Schwartz Machine Company, tells under what circumstances his company allows the operator to become involved:

> If we have faith in the operator, we let him edit and program. You want the operator to have a little input in the shop, then he feels like he's accomplished something. It takes some boredom away from the job. With CNC you're giving the operator a chance to run the machine even though the program comes from the front office.

If management does not have "faith" in the operator—the usual case—it becomes possible to exclude the worker totally from both editing and programming. A fairly prevalent attitude was expressed in the following exchange between engineers from Fairchild Aircraft and General Electric at the 1979 meeting of the Numerical Control Society:

> FAIRCHILD: It would be an advantage to make corrections at the machine site by programmers.
>
> G.E.: Yes, but we don't edit on the shop floor because we don't want the operator to do the editing.
>
> FAIRCHILD: I didn't mean that the operator should do the editing. Only the programmer.
>
> G.E.: It's really your choice. Where you want to do it and how you want to control it. The ability to do it is there with any of the new systems.

Ultimately, managerial prerogative in this area is symbolized by a key on the CNC control panel that locks it from unauthorized use. As an engineer from a machine tool company told me, "The key belongs to the man who owns the machine."

From a management point of view, the pressures of production often necessitate greater operator input regardless of the ideology involved. In other words, maximum control is sometimes incompatible with effective production. In these cases, real shop floor input is encouraged and the full capabilities of the technology come into play. We gain a glimpse of more productive and exciting alternatives and an idea of what other ways to structure production could look like. We have already seen how the pressures of a simultaneous plant startup and a new production run led to operators programming machine tools at Caterpillar Tractor. Some companies maintain as standard practice small work areas that are, in effect, islands of shop floor control within larger and more disciplined workplaces. If there is an unusual order or an emergency rework, these independent areas produce the parts without the layers of control, paperwork, and bureaucracy of the rest of the plant. These areas are not utopias but they indicate intriguing possibilities.

Two examples of this approach are the Cessna Aircraft plant in Wichita, Kansas, and a new British Overseas Corporation facility in England.[51] At Cessna the work area is only an MDI machining center and a drill press. When an emergency job comes in, the machinist produces the program and the part, and afterwards punches a permanent copy of the program on mylar tape. One typical rush job for 250 engine mounts was completed in two working days compared to about a three-week lead time using the conventional system. At British Overseas Corporation the independent shop is quite a bit larger, occupying about one third of the company's 100,000 square foot plant and using 55 machines. In this case, the autonomous groups seek to turn out small orders quickly without having to utilize the mass-production equipment in the rest of the plant. Although the machining centers have much slower cycle times, some work orders can be completed more rapidly than the time it takes to set up the production equipment.

Shop floor programming and editing also characterize some small shops where parts often have to be rushed to completion and there are few resources to cover for inefficient operation. In these cases, the social issues are frequently submerged because the machinist who is doing the programming is also the shop foreman or owner. One such shop is G and T Products in a northern suburb of Detroit, a small machine shop owned and run by a former machinist. The owner programs and edits three NC and CNC machines. He believes in running a tight ship and seeks to get by with the least expensive and usually least skilled workers he can find. But, his livelihood depends on turning out a quality product on time, and to survive in the cutthroat world of small job shops he needs a great deal of flexibility. If this means having workers do programming or editing on some jobs, he may not like it but he will do it, and then seek to maintain control over operations in other ways.

Ironically, these small shops sometimes show that CNC, which has the clear potential to expand the level of skills, can become the vehicle to operate with fewer skills. Consider Industrial Machine and Engineering, Inc., a small job shop in Linden, New Jersey. The plant employs under fifty people and is run by its owner, Steve Peti. The core of the shop consists of ten CNC turning centers. Peti is quick to point to his central problem before the introduction of CNC: Skilled workers were needed for shop operations. Each conventional machine required a competent machinist and there was a further need for ten to twelve highly skilled workers on setup. With CNC, Peti feels far less skills are needed on the machine tool. He contends that a machinist with one or two years of experience is easily the equal of a worker with fifteen years at the trade. Furthermore, one worker now operates two machines, which was impossible before. As far as setup goes, only two workers are required, and Peti's son does the programming. Peti goes as far as to say that if he had the capital for robots, he could dispense with many of the people at the machines altogether. Nevertheless, he allows some input in editing from operators on the floor, illustrating that increased input and skill are not necessarily the same thing. In the last analysis, he feels the machines have given him "better tolerances, faster production, and much tighter control over the shop."

In other cases, management may be willing to sacrifice increased productivity in the short run to solidify its authority in the long run. Leslie Nulty illustrates this in a case study of the fabrication shop for aircraft maintenance at the home base of a large domestic carrier.[52] The base itself is quite large by airline standards, employing 5,500 IAM members. Issues of productivity and managerial competence are particularly sensitive given the recent financial performance of the airlines. Several years prior to the study, the unionized members of the company voted to accept wage concessions when bankruptcy appeared to be a real possibility. Although an intermittent financial recovery occurred, further losses in 1980 and 1981 once again caused workers to lose a percentage of their pay.

The fabrication shop manufactures parts that are used in the maintenance and repair of the company's sizeable fleet of aircraft. It is staffed by highly skilled sheet metal mechanics, most of whom have considerable experience and are capable of operating all the machines in the shop. Traditionally, a worker is given an engineering drawing and then has complete responsibility to produce the required part.

In January 1980, an automatic turret punch press was introduced into the shop. This computer numerical control machine stamps batches of sheet metal parts and is capable of operating from preprogrammed computer tapes or from instructions that are entered at the machine. This programming option sparked a bitter struggle between workers and managers over who was to control the machine and therefore have authority over an important element of the work flow in the shop. The first signs of trouble began when the manufacturer had a demonstration of the new machine in the shop. Those workers who had previously bid to be part of the "machine crew" attended the demonstration and discovered that a free two-week training course was available. Desirous of learning the machine's programming technique and operating methods directly from the machine builder, the four or five workers involved requested to go to the school. To their utter amazement, "We were turned down, with the work manager telling us that we will be taught only what was required to operate the machine, and the school was only for programmers." Later, they learned that the airline had sent an engineer, a programmer, a foreman, and an electrical shop supervisor to the school. This was too much. A grievance was filed challenging the company's attempt to split off programming from their jobs as a violation of the contractual provision that gives the union jurisdiction over "all work involved in . . . overhauling, repairing, fabrication . . . and machine tool work in connection therewith." As the workers commented later, "The work program is a point of great concern because it is being used as a basis for justifying a removal of work such as flat pattern layout and template making, away from the sheet metal shop and into the hands of draftsmen and engineers."

They further maintained: "Something is wrong with a system that sends everyone involved with the equipment to school except the four people responsible to operate and maintain it eight hours a day. . . . A grievance was filed because we began to realize that we were being deliberately kept away from the information required to do our job."

The "machine crew," however, managed to obtain copies of the programming and operating instructions for the machine. For several weeks they pored over these manuals at home during the evening since they had no time for this study on the job. After this they began producing parts with their own manually developed programs. They maintained they were able to program, manufacture, and deliver a part more rapidly than it took merely to receive a workable tape from the programming department. In the first ten months of operation, they manufactured over 40,000 parts, averaging about forty-five minutes to produce a manual program and about thirty to forty-five minutes to store one into the machine's memory. They even requested a tape printer accessory so these manual programs could be recorded for future use. Moreover, the workers were able to locate errors in the manufacturer's manual that were only apparent to the programmers after "errata" corrections were later sent. The machine crew contended:

> Our methods have been quick and efficient and, up till now, we have been encouraged to develop our skills and work with our co-workers to help produce as many jobs as possible, because our shop is sorely short handed and this machine took money away from our department budget that could have been used to pay for as many as five sheet metal mechanics. We feel that this machine is different from other computer controlled machines on this base, because of its manual programmability features and that any sheet metal mechanic in the department is capable of learning the programming and operation procedures required to operate this machine. The mathematics required to figure out some of the more complicated programs is no harder than the setback and bend allowance figures we work with everyday.

The company, however, was less than pleased with this surge of productivity. Four months after the machine was brought in and while the grievance was still being discussed, an on-off switch was installed to lock out the manual programming features of the CNC unit. Given the skills of the shop floor workers, this proved to be more a symbolic statement about who controls the machine than a real roadblock to the continual production of programs by the "machine crew." Nevertheless, the airline posted an elaborate twelve-step, three-page summary of procedural rules for "requesting, procuring, and implementation of numerical control part programs." Under this plan, the shop must ask for and receive a computer program from the programming group prior to any work order being filled. Rules 8 and 9 give the programmer clear authority over machine operation during the trial tests of the tape.

8. N/C Programmer will be present and have control of operations during performance runoff.

    - Regular machine operator will perform any tool set-up changes at instructions of N/C Programmer.
    - Regular machine operator will operate machine during performance runoff.
    - Any N/C program editing required will be accomplished by the N/C Programmer.

9. Based on N/C program editing required during performance runoff, N/C Programmer will revise N/C program as necessary.

The system was made even more cumbersome by the fact that the programmer's offices were at one end of the airfield and the fabrication shop was at the other end. The result of this heavy overlay of bureaucracy on the system was a turnaround time of a week or more on most jobs. Moreover, the workers were really taken aback to learn that the convoluted nature of the operation had so overloaded the programming department that it was contracting work out just to stay afloat. As one member of the machine crew put it:

> I and my fellow workers have been harassed and frustrated because we dared question a policy that doesn't work, won't work, and concerns work at which we are paid professionals. If a policy to do sheet metal work was to be set up and put into operation, it would seem obvious to me that the people who would ultimately control and accomplish the work should set rules, not a slung together programming department on the other side of the field.

What the skilled workers proposed instead appeared to be a model of production efficiency. The programming department would be retained to provide specialized support services and to maintain permanent records of the programs that are used. Whenever possible the workers on the floor would program the machine or improve existing programs, responsibilities that are analogous to the planning a skilled metalworker does on conventional equipment. If any unusual problems arise, the programming department would be consulted or if necessary the job would be contracted out. In their open letter, the machine crew stated their overall concerns.

> Our motives are simple ones of profit and production with job security in return. With part of our pay going to support this company we feel responsible to do our best to stop unprofitable practices whenever possible. . . .
>
> We have shown our willingness and ability to learn the operation of this machine and proven we are capable of doing all we say we can.

While these would seem more properly to be managerial concerns, in this case issues of control and authority were paramount. The company sought to use a new machine as a vehicle to transfer work out of the bargaining unit—from unionized sheet metal mechanics to nonunion programmers—and to reinforce its command over production decisions. In order to do this, a price was paid in lost productivity. A further irony is that the airline has strongly urged more labor-management cooperation and has solicited more productivity-raising suggestions from workers.

Another case in which control superseded productivity as a managerial concern was apparent on a visit to an aerospace plant in England. In this modern facility, NC and CNC machine tools provide the backbone for jet engine production. At the time of my visit, a running battle was going on between machinists and programmers over who was to edit the tapes on the CNC controls. The machinists were demanding the right to do their own editing. Because they were paid on a group bonus, they felt both their output and their earnings would increase if they had more control over the machine. The machinists maintained that the programmers were largely recruited from engineering ranks and as a result had little actual machining knowledge, a serious handicap in trying to optimize part programs. One machinist complained that a job had required over twenty reworks of the program, which became spread out over a three-week period because the programmers were busy with something else. Another machinist was frustrated because tapes were only edited on the day shift, handicapping work done on other shifts. The programmers, on the other hand, were insisting that the editing of tapes was part of *their* job, which they were fully competent to do. To enforce this, they "blacked" the CNC machine over which the dispute had arisen, refusing to prepare programs for it until management isolated the edit facility from the machinists. (The situation was further complicated because both the machinists and programmers were members of different divisions of the same union.)

Some machinists, however, had learned how to do substantial alterations of the tapes and were doing them surreptitiously whenever it made their job easier or increased their group bonus. One worker on a less sophisticated machine had even obtained a tape punch and was secretly preparing his own tapes to increase his production. His immediate supervisor knew about the practice but was reluctant to intervene because of the fear of interfering with this new source of productivity.

Top management, however, was siding with the programmers in order to preserve its authority on the shop floor, realizing that the ability of machinists to produce more also might mean the ability to produce less. Productivity was sacrificed in the short run for the chance of predictable output in the long run.

The managerial drive for more effective shop floor control and the flexibility of computer numerical control may seem to contradict each other. In other words, if management is so interested in extending its authority, how did a technology that potentially undermines centralized control come into such widespread use? There are a number of explanations to this apparent conflict. First, although CNC lays the basis for wider decision-making and power on the shop floor, it can also be a vehicle for expanded managerial jurisdiction. In reality, there is now a choice as to precisely how much decision-making should be delegated to the shop floor. Second, once a program is prepared and optimized, the machinist is outside of the control loop in any case. Whether it is NC or CNC, the knowledge, experience, creativity, and talent of the machinist are embodied in the program. Finally, the drive for control is not the only managerial motivation in the workplace. In particular, CNC makes possible an improved and more flexible machining process. Part of the price to obtain these benefits may be blurred lines of command and some erosion of authority.

New technologies that increase shop floor input, however, generally create little interest among machine-tool buyers. A new programming method that was developed by David Gossard, a professor of mechanical engineering at MIT, follows this pattern. Gossard's approach permits a skilled machinist to produce NC tapes with little formal training through what he calls "part programming by doing" or analogic part programming. Gossard sought to simplify part programming in order to involve machinists in the process, and thereby speed the diffusion of NC into small and medium-sized shops. In the system, a drawing of the workpiece appears on a cathode-ray tube, and the machinist uses a joy stick to trace the cutter path with a simulated tool in a way that is reminiscent of computer games in amusement arcades. The cutter path is then recorded and becomes the part program. Gossard built a prototype of the system, and in initial trials, machinists without programming experience were able to program moderately complicated parts after a few hours of practice.

Gossard felt that his system would free the machinist from artificial constraints in preparing machine instructions. In other words, the machinist could directly translate experience-based knowledge into an abstract part program.

> By providing a mechanism whereby the information regarding a cut is conveyed in a manner resembling, as closely as possible, the conceptual process of the machinist, the constraints imposed by symbolic part programming are largely avoided. As a result, the complexity of the part programming task is significantly reduced.[53]

While Gossard's system received some publicity in the trade press at the time of its introduction, no major company was interested enough to develop it. In a 1980 trip to Japan, however, Gossard was warmly received by a Japanese company that was experimenting with a system remarkably similar to his prototype.

Ultimately, important elements of what a part programmer does will be automated and, in an increasing number of cases, programming itself could be eliminated as a separate occupation. This underscores the dynamic nature and extraordinary range of computer-based automation. Not only is the entire production process affected, but even those jobs created as a result of automation today may be automated tomorrow. The same fracturing of skills and control that machinists have already experienced are now being brought to the programming department. A report done for the Machine Tool Task Force predicts that the trend is to fold programming into the design process itself "eliminating the programmer as we know him today."[54]

The key to automating programming is computer-aided design (CAD), the linking of the computer to the design process through TV-like terminals with which an engineer interacts. In this process, which will be discussed in more detail in chapter 6, an engineer sitting in front of the screen uses an electronic pencil to design a part. As the part is drawn, computer systems automatically generate the path for an NC cutting tool, eliminating one of the programmer's central tasks. One such system is CAD-D, an advanced design approach developed by McDonnell Douglas. Even when the APT language continues to be used, new systems such as regional milling allow a programmer just to define the boundaries of a particular surface, say a mold cavity, and the program is automatically developed that will carve out that surface.

Systems such as CAD-D do away with the need for many traditional manufacturing skills between design and production. An amplification of this system, called Fast Cut, goes even further and lessens the need for skills on the shop floor. Fast Cut utilizes CAD-D methods to design tooling, the fixtures that precisely position aircraft components while they are being machined or assembled. The underlying principle is building a tool by bolting together relatively simple two-dimensional shapes, more or less as one would use an erector set. At the design stage, a picture of the aircraft part is called up on the computer screen, the tooling is designed to hold the part, and then the Fast Cut program automatically generates a pattern that will allow all the tooling shapes to be bolted precisely together. The program insures that there will be a common set of holes between any two parts that are supposed to mate. Additionally, an NC program is automatically created to machine each individual tooling component. All that is left for the designer is to select a number of parameters concerning details, such as cutter speed, from a "menu" of choices that appear on the screen. According to Jack Jessen, a tooling manager at McDonnell Douglas in Long Beach, California, "This eliminates the programmer completely. We taught our tool designers to use this and they don't know the first thing about programming."

On the shop floor, a highly skilled toolmaker is no longer needed to machine and assemble these fixtures. Instead, the requirement is for what Jessen refers to as a "Tinkertoy assembler," someone who bolts together pre-cut and pre-marked details.

There are, however, some limits to the use of systems such as Fast Cut, both technical and social. One technical limit is that some aircraft tools are too large to be done this way. On a fixture that must hold a large wing, for example, it is possible to design the entire jig by computer but difficult to find an NC machine large enough to fabricate it. This type of project, however, can be constructed by using smaller subassemblies that were developed utilizing Fast Cut. The total labor time might be cut by 30 or 40 percent compared to conventional tooling methods, but somewhat more skill is involved in the assembly. Also, there are some more complex types of fixturing that still require a toolmaker. Nevertheless, about 50 percent of all assembly tooling at McDonnell Douglas in Long Beach is now built using Fast Cut, and this could easily rise to 75 percent by 1985. What the future might hold, according to Jessen, is a "widening ratio between Tinkertoy assemblers and toolmakers on the shop floor."

An important social limit, at least in the short term, is the resistance of people whose skills are being eroded or destroyed. Jessen maintains:

> how successful any technical application is depends on how much cooperation you get from everybody involved.

> Now, the natural tendency on the part of anybody in that chain from design to completed tool that feels threatened by what he sees is to drag his feet a little bit, that's human nature. So, you have problems along that line that have to be solved any time you get into technologies that are going to eliminate certain skills or trades.

The formal elimination of a classification such as toolmaker would require the consent of the union. Significantly, the experience of NC is pointed to as a model for how to deal with the removal of skill from other occupations. Jessen elaborates:

> The union is naturally going to want to take a hard look before they go and give up any of their skills. This is a bridge that we've crossed in other areas and we've eventually convinced the union that you don't need to have the skills that you used to have. For example, an NC machine operator doesn't need the skill of a full-blown machinist who knows how to make all his own setups and read all of these dials and everything. So the NC operator has been widely accepted in industry and I think that eventually the use of less-skilled labor for assembly tooling will also do that. Unfortunately, we're not there yet.

Some observers are concerned about this trend. Eliminating the part programmer removes production decisions even further from the point where metal is actually cut. This assumes that all the complexities of the machining process can be unraveled, quantified, and modeled into an automatic series of actions. It would be impressive to be able to do this at all, but to do it in an optimal way is far more difficult. The stumbling block is not automating the cutter path, which has already been done, but the selection of cutting conditions. As Jessen puts it:

> How do you develop a software package that will take into consideration a whole multitude of different situations and conditions, meaning what kind of metal is involved, say stainless, titanium, aluminum, cast iron, or whatever, or the different methods of forging, or any of a number of other parameters. I don't see a program like that in the near future at all.

Seeking to bypass human input completely may create a level of complexity and frailty in the production process that is limiting at best and totally unmanageable at worst. As another McDonnell Douglas manager put it, "Part programming is not strictly a science but more of an art or a craft, and it is very difficult to automate a craft." Some astute production managers realize that this may be the wrong road to be going down. As Norm Hopwood, a computer manager at Ford, put it, "I don't know why anyone would want to eliminate the part programmer. He's the best person we've got." Other managers are concerned that by eliminating decision-making from programming, the same boredom that affects NC operation will set in. Bob Nordhuff from Douglas Aircraft warns, "When you put a person on an NC machine, people start goofing off. The machinist doesn't have to think that much on NC. What's happened down there, I don't want to happen in the programming office."

Pushing programming further and further from where production takes place has other dangers as well. At the present level of technological development, the assumption that the "feel" of metal cutting is no longer of value is an illusion, heightened by dreams of total shop floor control. These dreams lead to a use of programming that widens the gulf between conception and execution, between the shop floor and the engineering office, and between the hand and the brain. Instead, the technical possibility now exists, in many instances, for the worker to fully program the machine tool. In other cases, a separate programming function is technically necessary, but even here the machinist could play an important role in selecting the cutting conditions or improving the program. Moreover, new forms of work organization could be designed to include the machinist more fully in the process. One way would be to rotate the machinist through the programmer's job, thus giving workers programming experience and the hands-on feel of production to engineering. The value of this is illustrated by the wider role for machinists that production pressures sometimes bring about.

Once a program is prepared and optimized, however, the machinist is largely out of the machine control loop in any case. More than an input into programming is therefore required for the worker's position in production not to erode. The real issue is power over the way manufacturing is carried out, and computerization does not have to undermine the worker's position. In conventional metalworking, for example, the interplay between a prototype machinist and an engineer gives the worker substantial input into design. Replicating this with computer technology might involve providing the machinist with access to a design terminal while the engineering work is still taking place. Moreover, why should the goals of computerization be limited to retaining the workers' current position in the production hierarchy? New possibilities exist to expand human participation. The result might not only be an improved quality of life on the job but increased productivity as well.

THE IMPACT ON THE MACHINIST

How do machinists feel about numerical control? The answer can vary widely depending on the nature of a machinist's experiences and attitudes. Underlying these specific considerations, however, is a powerful and pervasive ideological theme: Technological change is always progressive and inevitable. Consider, for example, the metaphors of technological change. When it is not being compared to an onrushing locomotive, it is likened to a force of nature, say a tidal wave. Anything destroyed in the path of technological development is presented as the necessary cost of achieving a better future. Resistance is painted as futile and the only real alternative is to get out of the way. Such thinking precludes any serious discussion of alternatives.

This ideology of progress is a powerful though hidden factor in shaping the attitudes of workers toward numerical control. All of the machinists I interviewed said at one point or another in the conversation something along the lines of "you can't stop progress." I spoke with the president of a large aerospace union local who complained bitterly that the local had been reduced from 24,000 to 9,600 jobs in ten years, the apprenticeship program had been eliminated because the company felt that fewer skills were needed, and machinists were now compelled to operate two or more machines. After detailing these grievances, he concluded the conversation by saying, "Well, it's inevitable in any case." This feeling of powerlessness in the face of change undermines the value of experience and judgment as a guide to influencing action. We begin to see ourselves only in the role of the object rather than the subject of change.

Moreover, technological development in general seems shrouded in mystery and this is reflected in the reaction to new technology in the machine shop. For older workers, the use of computers arouses acute attacks of anxiety. Seasoned diemakers in one shop that I visited became so intimidated by the flashing lights and display screen on the control panel of a new computer numerical control machine that they were reluctant to operate it, even though far less skill was involved than what they normally did.

Younger workers who have grown up surrounded by

computers may be less intimidated but often seek an individual rather than a collective solution to the problems of new technology. Since influencing the direction of development seems impossible, the most capable and ambitious workers seek the best personal positions they can within the existing framework. Rather than striving to broaden what a machinist does on the shop floor, a talented young worker may instead simply choose to become a programmer. With a rapidly developing technology, "getting in on the ground floor" is a potent incentive.

Nonetheless, the ideology of progress, although it helps form the wider view that workers have and mediates what they say about technology, does not paralyze direct action against managerial designs. There can be profound contradictions between the acceptance of numerical control theoretically and the struggle against managerial directives for its use. One by-product of this is considerable confusion among some observers of the workplace who confuse the answers to questionnaires and the responses to interviews with shop floor reality.

Other factors also mold the attitudes of machinists to numerical control. The frame of reference of the worker—an amalgam of personal experiences and feelings about the trade—can be central in developing a point of view about NC. In highly skilled one-of-a-kind prototypes or limited production shops, there are often strong personal feelings about craftsmanship, and the devaluing of skill can cause deep resentment. In England, I spoke to a worker in an aerospace plant who had been a highly skilled machinist for seventeen years and who truly enjoyed his work. For the six months prior to my talk with him, he had been running an NC lathe and felt a deep frustration. He movingly stated his plight:

> I've worked at this trade for seventeen years. The knowledge is still in my head, the skill is still in my hands, but

> there is no use for either one now. I go home and I feel frustrated, like I haven't done anything. As a result, I find myself wanting to make things around the house. I feel something has been taken away from me that I could put into the job.

These attitudes were echoed by some workers at the McDonnell Douglas plant in St. Louis. Realizing the necessity of computer technology for advanced aerospace production, these workers are nonetheless opposed to the way numerical control is used. As we have seen, responsibility on the job has increased while skill and training have been substantially reduced. One machinist complained:

> The machinist trade is becoming less and less here because of automation. For a young man, it's a poor occupation to get into. Eventually they'll phase it out altogether. They'll have a closed circuit TV and one man running six machines. If they could hire a smart monkey for a few peanuts they would.

Ben Dubraski, the late assistant directing business representative of the local, was quite direct in his response. "Automation is killing us here. They no longer need the skilled people. Those machines are built to run a long time and they'll outlast us."

Chris Laverty, a diemaker and a UAW committeeman at the huge Oldsmobile complex in Lansing, Michigan, is also very critical about the ways in which numerical control is being used. In a paper he wrote on NC, he commented:

> The job has been specialized, trivialized, and down-graded. Merely pushing buttons and watching warning lights is unlikely to hold intrinsic interest and challenge for very

> long. Initiative is no longer required and the work is boring.
>
> Pride of craftsmanship has been destroyed for the man who operates a Numerical Control machine tool. The machinist can no longer identify with the product. He used to make the part from start to finish and received a lot of satisfaction from it. The job has been routinized and bureaucratized and has become less and less interesting.[55]

Laverty interviewed a number of his colleagues and found frequent complaints about stress and frustration. One NC operator remarked: "On my old job I controlled the machine. On my present job, the machine controls me."[56] Another machinist complained about the anxiety of standing in front of an experimental machine whose cycle you had no hand in planning. For him, it was obviously more than a mythical fear.

> I accidentally pushed the wrong button and things flew all over the place. The machine took off at four hundred inches per minute instantly scrapping the part, the fixture, and the spindle on the machine. I didn't get hurt, but it cost the company $13,000 to fix the machine.[57]

This anxiety is widely shared by numerical control operators. Greg Trapoti, a machinist in the die room at the Rouge, comments:

> When you're running on tape, it is no longer in your hands and it can make you very nervous. If the programmer had a bad night, and forgets to allow for a rib in the casting, the whole part can go flying off the angle plate and into the aisle.

Experience such as this led another worker to comment:

> On my old job, my muscles got tired. I went home and rested and then I was O.K. On this new NC mill, your muscles don't get tired but there is mental strain long after you have gone home.[58]

Laverty sums up by saying that "night work, increased boredom, and the vigorous industrial discipline in general often make these men temperamental and difficult."[59]

Stress and boredom are recurrent results of NC. A 1970 article in *American Machinist* indicates that these problems have been around throughout the history of this technology.

> A major problem we have faced in the past in manning our NC machine tools has been the apparent fact that few skilled, experienced machinists long are happy with continuous assignment to the operation of an NC unit. They soon feel that their hard-earned skills are being wasted, and they become bored. As a result, we have experienced a complete turnover of NC operators at least every twenty-four months.[60]

Rather than modifying the technology to address the problem, the author presents another solution. He recommends that only new workers be assigned to NC. For those that last long enough to make it through a two-year training program in computers, the reward will be a promotion to part programmer.

Boredom can lead to occupational hazard. The director of the Machine Tool Industry Research Association in England summarizes a recent report on NC by stating:

> The operation of some NC machine tools has already become mere machine minding and operator concentra-

> tion is difficult to maintain. Boredom can lead to loss of concentration and thus to increased risk of accident.[61]

Sometimes it is the supervisors who suffer negative effects. One worker maintained that numerical control was a fine technology as far as he was concerned but that his foreman had become exceptionally edgy since the new machines had been introduced. The supervisor's problems stemmed from the fact that he had never actually operated an NC machine himself and was therefore uncertain about the capabilities of the process. This uncertainty was compounded by extreme pressure from upper management to get the maximum production from the new more capital-intensive equipment. This is a far cry from some early promises that numerical control would eliminate the "human element" from supervision.

> Supervision can do very little to alter or check the way the machine is operating. . . . The supervisor's principal job becomes one of work-flow expediting. . . . He is more a manager of machines than a manager of men.[62]

Some skilled workers are more ambivalent about numerical control, seeing some benefits but feeling apprehensive about its effects on their trade. One person who shares this viewpoint is Jim Daley, a diemaker for a number of years in the Rouge plant and now a parts programmer. Sitting in a bar across the street from the Rouge plant after work one day, he commented about numerical control:

> Most people are glad to have it because it eliminates bull work and dirty work that machines are supposed to do, but it is breaking down the trade and that's bad. I could make my kid a die-maker in a month. It's only one step away from the assembly line.

Other tradespeople are not particularly threatened by the loss of skill requirements. If NC is a relatively small part of the overall machining operation and workers rotate on and off the automated equipment, they may view operating an NC machine as a welcome break. In one case, a diemaker at the Rouge had considerable pride in his craft, but total contempt for the company and his working environment. His comment about numerical control was simply, "I don't come to work to have a good time." Confident in his skill, the less he had to do at work the better he felt.

Since NC affects so many widely varying types of machining, the specifics of how it changes job duties and work relations in a given situation influences worker attitudes. In some medium-run production applications, for example, the actual skill levels on the job are only minimally affected by the new technology since NC replaces machines that did not require a high skill level in the first place. The response of workers therefore is more related to other factors. Sometimes worker reaction is positive. In a tractor-components plant, a machinist was favorably impressed by some of the physical characteristics of NC.

> It's spoiled me rotten. You need limited tools, the machine is accurate, the noise level is almost none, you don't have problems with chips, and you don't even get your hands dirty.

Mike Reicher, a chief steward at Caterpillar Tractor in East Peoria, Illinois, agrees:

> NC is a lot less hassle than a regular machine. I love it. You have the advantage of cycle time, less crankiness, and it doesn't take a lot of time to change size.

Another machinist prefers working with NC because it gives him more time for himself. "When the machine is working, I am not."

Some workers in medium-run production have a far more negative response, even though the change in job content may be minimal. One worker complained that management was able to monitor the job more tightly as a result of NC. Another worker felt frozen out of the production process with the new technology. He complained: "It hurts seeing how they try to do some jobs. You want to say something, but they don't want to listen. They have their program and that's it." Some machinists complained about increased noise because the machines are running faster and others were concerned about being more isolated since the machines are larger and spread further apart.

The way NC is introduced also affects worker response. Workers' fears were initially assuaged by the notorious reliability problems of the early generations of machines. At the outset, machine tool builders and salesmen gave managers visions of new machines coming in one door and skilled workers exiting at another, and these feelings were, of course, communicated to the skilled workers. Not only didn't the technology of the 1950s and 1960s live up to these extravagant promises, but very often the machines themselves didn't work. The rate of introduction is another important factor. Introducing one or two new machines at a time may arouse only curiosity, but a rapid influx of new technology is likely to spark a much sharper reaction. Since NC has generally been introduced rather slowly, there has been a more subdued response than otherwise might have been expected.

The impact of NC on employment has often sparked controversy among skilled workers. There tend to be two very different responses. Some skilled machinists feel that the increased speed and capability of computerized machining will mean that more products can be produced with fewer work-

ers, thus eroding employment. Other workers, however, are convinced that these superior performance characteristics are vital for a shop to be competitive, and therefore firms that use NC will win work from less-advanced enterprises, thus preserving jobs. Moreover, NC sometimes lays the basis for an industrial restructuring that affects workers' attitudes concerning employment. In the tool and die industry in Detroit, for example, the introduction of NC technology caused an increase in the concentration of the industry because only the largest independent shops and the captive shops owned by the automakers could afford the more sophisticated versions of the new equipment. As a result, machinists I spoke with in the hard hit small shops viewed NC as more threatening to their jobs than workers in the larger firms whose shops may have become busier as a result of NC.

The fear of job loss has been allayed in the past by the introduction of NC technology in good economic times when it generally has not resulted in the direct displacement of workers. (During recessionary periods, the capital for new equipment is generally not available.) The authors of a study done at the Massachusetts Institute of Technology expressed surprise at how little resistance there was to the introduction of NC in twenty-four small- and medium-sized shops in the eastern and midwestern United States. The study found:

> The firms we interviewed had experienced no resistance to the introduction of NC from their workers. This result may appear unexpected, in view of the greater productivity attributed to NC machines. On closer examination of this subject, however, it is clear that NC machines made their appearance during a period of growth for most of the firms involved in the study. . . . It is possible that worker attitude may change in the future, but the tradition of having Numerical Control machines in the shop is already well established.[63]

The Computerized Factory: On the Shop Floor

NUMERICAL CONTROL, with all its far-reaching capabilities, is only the first step toward automating production. While NC brings the cutter on the machine under more direct managerial control, it leaves untouched all the operations involved when the part is not being cut: operations such as loading and unloading workpieces, transferring parts between machines, scheduling production, maintenance, and quality control. Consequently, the next steps toward computerization target the activities of the remaining workers on the shop floor and the support services necessary for production. Or, as one manager at Cincinnati Milacron put it, "non-machining time is the issue in all of metal cutting."[1] In fact, the success of NC in automating metalcutting has been both a spur and a model for further computerization. According to a report from the School of Industrial Engineering at Purdue University:

> Analysis of the activities involved in keeping an NC machine tool cutting metal for its entire running time shows

> that human involvement in the actual production process is the primary slowing down factor. The solution to this problem, suggested several years ago by a number of independent studies, is to automate as many as possible of the production functions in the same manner as the NC cutting process.[2]

The rest of production is being computerized in two related ways: first, the automation of specific operations, and second, the tying together of these "islands" of automation into larger systems. An example of the first approach is the way NC automates machining or the way robots eliminate workers in the loading and unloading of machines. An example of the second is the same computer directing both the operation of an NC machine and a robot loading and unloading it. Ultimately, the goal is a hierarchy of computers directing production. Moreover, this super computer system will define, collect, and control the processing of information throughout the factory. Many jobs are of course eliminated, but as significant is the degree of command over the activities of the workers who remain.

Out of the explosion of new technologies available, I will focus on three: flexible manufacturing systems (FMS), an attempt to fully automate batch production by carrying numerical control to its logical extreme; robots, an incredibly versatile technology that can be used throughout the factory; and management information systems (MIS), a computer network that collects data about production while the event is taking place. I will also look at the introduction of a very ambitious computer scheduling system in batch production and its subsequent collapse, a case that shows the importance of the human response in the workplace and underscores the fact that the success of computer systems is hardly preordained.

These technologies represent some very different technical applications. The diversity is further enhanced by the enormous range of production operations and industries in which they are used. Flexible manufacturing systems, for example, are exclusively deployed in small- and medium-volume production. In contrast, robots and management information systems are used in mass production as well as batch production. Together, these systems produce products from washing machines to jet aircraft engines. Moreover, very different managerial strategies govern their implementation and use. Nonetheless, some common social purposes influence their development and operation and this chapter will delineate these themes. It will look at the interaction of workers, managers, and new technology at the point of production and, in particular, the ways in which the drive toward managerial control plays itself out in the context of the factory as a whole.

FLEXIBLE MANUFACTURING SYSTEMS

The White-Sundstrand Machine Tool Company knows how to attract the attention of a managerial audience. In a full-page ad in *American Machinist*, they show an obviously worried metalworking executive. Holding the phone with one hand and gripping the side of his face with the other, he is grimly gazing out over a mound of computer printouts and reports scattered over his desk, with an untouched cup of coffee sitting nearby. Across the top of the page, large block letters ask the question "LOSING CONTROL?" Since the answer is obvious, the ad continues with a solution: "REGAIN COMMAND WITH AN OMNICONTROL DNC SYSTEM." What is DNC? It stands for direct numerical control and describes a system in which individual NC machine tools are linked to a central computer. It is possible to store a program for every part that is made in the shop, and these programs can then be easily transferred to any machine in the network. Moreover, the system is designed to receive feedback on how the machining activities are being carried out. The ad continues:

> The computer-aided manufacturing system creates a vital information loop. You not only control what your machines do, but you also receive a steady flow of information back from the shop floor. Having constant awareness of production status greatly strengthens your control position.
>
> It also allows you to respond quickly to program changes, part prove-out, and program optimization with source level editing. Decisions on better utilization of machines and personnel are made much easier. Costly downtime and harassment from tape-related problems are eliminated. Control is reestablished and the pressure eased.[3]

Some of the feedback concerns technical aspects of the machining process: items such as monitoring electrical and mechanical failures, broken tools, or a lack of materials and, if necessary, shutting the machine down. Other aspects of the feedback, however, are designed to control many of the activities, and eliminate some of the control tactics, of the worker that NC itself may have missed. A DNC system, for example, can be set up to record when the override switch is used and for how long, printing a record that is later evaluated by management. At Lockheed-Georgia, there are plans to install TV screens next to the machines. These screens will display what Joseph Tulkoff, director of manufacturing technology, calls "black book information—a body of information, usually submerged, that has been previously learned about a job." Information such as who ran the machine, on what job, what speeds and feeds were used, and what the tooling setup looked like. There will also be, according to Tulkoff, "shop aides, little innovations, cautions, special notes concerning quality."[4] The pioneering study for managers, *Computer Integrated Manufacturing*, details an awareness of the limits of NC in this regard and the value of direct feedback from the shop floor:

> One company had been plagued by a series of NC machine stoppages charged to dull tools. When the operator wanted a break from his routine he would decide that the tool was dull, stop the machine, remove the tool, and walk clear across the shop to the toolroom where he would apply for and receive a similar but freshly sharpened tool. He would then proceed back to his machine and install it. Detours past the coffee machine were not infrequent.

The solution: the use of direct numerical control, instead of the worker, to make the decisions about dull tools.

> When the DNC connection was set up, any detection of a dull or broken tool was immediately signaled to the toolroom by the computer link, together with the location and number of the tool that had been in use at that moment. As an automatic procedure, the toolroom foreman would send a messenger immediately with a replacement tool. When this system became established, the number of dull tool complaints dropped by 70 percent.[5]

This sort of a setup can result in some significant productivity increases. A DNC network at the giant GE jet engine plant in Evandale, Ohio, for example, increased productivity by 15 percent over what was already viewed as an extremely efficient area. The system linked over 100 NC machines in the rotating parts group.

A computer directing the operation of machines is only the beginning. The ultimate approach to numerical control machining is the flexible manufacturing system (FMS), a complex of NC machine tools, automatic shuttles to move parts between them, and centralized computer control for the entire operation. The flexibility has two aspects. The first is the ability of NC machines to handle a range of different parts and easily to make modifications to existing parts. The second is the ability to move parts between machines in varying order depending on production needs. Theoretically, at least, raw materials are loaded onto the carts at one end and finished parts come out at the other with very little human intervention in between.

In operation, these systems are impressive. At the Messerschmitt-Bolkow-Blohm factory in Augsburg, West Germany, fuselage parts are machined for the Tornado fighter aircraft.[6] The largest machines are eleven mammoth gantry mills, capable of machining three to six identical parts at one time and of holding pieces up to fifty feet long. Self-propelled bright blue trolleys carry parts to these giants or any one of thirteen other numerically controlled machines, following wires embedded in the floor. After dropping off their load, they pick up completed jobs, which are then taken either to other machines or out of the system. Meanwhile, hundreds of cutting tools are passing by overhead on their way to the machines. Hanging on trolleys suspended from fixed tracks, 15,000 are delivered monthly. Off the shop floor, two computers coordinate management operations such as work scheduling, inventory, and NC programming, while three more computers are each dedicated to a production subsystem: the machine tools, the automated tool storage system, and the automated transport of parts.

While certainly fascinating to watch, an unanswered question is whether these systems represent a production advance or a factory level version of the Concorde, a tribute to technical virtuosity in the service of some misguided social goals. After all, the Concorde is a superb technical achievement if the parameters are the absolute necessity of getting from Washington to London in three hours, lots of readily available fuel, and largely unlimited operating costs. As far as an FMS goes, does the design of the system reflect technical necessity or the managerial obsession to eliminate worker input at virtually any cost? Now the possibility, of course, exists to design these systems in a way that allows substantial human judgment to be exercised. But, what is more likely is a technical onslaught against any worker participation. The consequence of this onslaught is an unnecessarily complex design, top heavy with layers upon layers of every conceivable kind of hardware and software to avoid any worker input at all. In some cases, this mentality leads to the installation of an FMS when a far simpler approach, such as individual machine tools, would do. In other words, the heavy artillery is dragged into position when all that is necessary are a few rifles. The first casualty of this approach can be productivity itself, but even when efficiency is maximized it may come at the destruction of the work environment.

Managers seek different things from these systems. As usual, there is a combination of technical and social goals. Technically they are targeted to fill the gap between standalone NC and highly automated transfer lines. As we have seen, transfer lines are capable of producing millions of the same part provided the design doesn't change, while NC can accommodate innumerable design changes but only by sacrificing some of the economies of volume manufacturing. In the range of mid-volume production, say 2,000 to 30,000 parts a year, flexible manufacturing potentially offers some of the benefits of both approaches.

One leading benefit is a lessening of direct labor. A considerable number of jobs can be eliminated through the automatic transfer of parts between NC machines. Beyond this, Brian Moriarity, a former project manager on FMS at the Charles Stark Draper Laboratories, a leading research center for automated technologies, maintains that labor can be minimized by keeping all the machines busy through full scheduling control. According to Moriarity, "In a job shop, you have a lot of playing around to do to keep all the machines busy but with these systems it's all built into the software."[7] This view is echoed by Thomas P. Shifo, a vice president of White-Sundstrand. He maintains that: "The utilization of stand-alone machines—even NC machines—is normally less than 50 percent mainly because so many decisions have to be made about where the parts go and what is the optimum use of machines and manpower at any given time."[8] Shifo asserts that the utilization of an FMS can go as high as 70–80 percent.

The technical and economic reasons for adopting FMS are explored in some detail in a series of case studies done by Donald Gerwin, a professor at the School of Business Administration at the University of Wisconsin-Milwaukee, and Jean Claude Torendeau, who is affiliated with the Institute for Social Research in Industry at the Norwegian Technical University. These researchers studied four firms, each located in a different country—the United States, Britain, West Germany, and France. The motivations that led each firm in the direction of an FMS are somewhat different but there are some common themes. The drive to reduce direct labor is pervasive. Additionally, in three of the four companies, a new product line and the requirement for more productive capacity was the starting point. In the U.S. case, the need for a modernization of facilities was also important, and with the British firm this was combined with a new management pushing for technological change.[9]

The differences are related to the specific context in which each firm finds itself. The U.S. firm, for example, which specializes in making tractors, is a division of a large diversified manufacturing corporation. The division has had an FMS since the early 1970s and, in fact, was the second U.S. firm to adopt one. A transfer line might have been a good bet to produce a new product line, but in the highly competitive tractor industry, uncertain market forecasts, which had a tendency to be revised downward, created a sense of risk. Moreover, these new forecasts lowered the projected volume below a point where a transfer line would have been feasible. The central goal became minimizing risk rather than optimizing production. Under these circumstances, the flexibility of an FMS became a key reason for its adoption.

The British firm, also a subsidiary of a larger corporation, produces electrical motors and generators. A new management felt that a direct numerical control system could double output per man-hour and thereby increase the company's competitiveness. The firm also sought an improvement in its manufacturing delivery time. Rather than spending two weeks getting important parts through the shop, the target became twenty-four hours. Since even with conventional methods only ten hours were spent in actual machining, most of the improvement would come from systemization. In addition, the DNC system promised to provide management with a centralized source of information about shop activities, thus increasing manufacturing control. Finally, the company maintained that a skilled labor shortage in Britain made the requirement for fifty fewer workers an advantage.

The German company is a manufacturer of commercial and military aircraft. Two central considerations in choosing new technology were maximizing machine utilization and reducing the number of workers required. The company sought to minimize the number of workers (aside from immediately reducing its direct labor costs) because it planned only about ten years of continuous production on the new aircraft in question and the combined pressure of strong unions and Germany's codetermination law provide important constraints on layoffs.

The French company, a manufacturer of industrial vehicles, sought improvements in delivery time, inventory reductions, and a better utilization rate for its machinery. The flexibility of FMS was also appealing in an industry that is highly competitive and cyclical. The company felt that product changes could more easily be made and, in a severe downturn, the machinery might have other uses.

Underlying these various technical and economic objectives are some broad social goals concerning control of the workplace. While sometimes not directly stated, they are central to many of the strategic choices that are made and are especially important since FMS procurement decisions, like NC, are seldom fully spelled out in financial terms alone. As Paul R. Haas, a vice president in charge of Kearney and Trecker's Special Products Division, the leading manufacturer of these systems, puts it: "The flexible manufacturing system is really more of a management tool than it is a manufacturing tool. There is nothing in this technology that allows you to make parts you couldn't make before."[10]

Iron Age magazine elaborates the feelings of many metalworking executives about the value of FMS as a management tool:

> Labor's role in manufacturing, particularly as regards control over production rates and product quality, is being thoroughly reexamined. Workers and their unions have too much say in manufacturing's destiny, many metalworking executives feel, and large, sophisticated FMS's can help wrest some of that control away from labor and put it back in the hands of management, where they feel it belongs.[11]

With this in mind, technological choices in the design and development of an FMS tend toward making workers as peripheral to the system as possible, thereby reducing their leverage on the job. Paul S. Borzcik, a manager of technology applications at TRW, discusses the role of flexible manufacturing in breaking the operator's control over production speed:

> FMS's may offer the ability to sever the human operator from the operations, initially for a few hours and eventually for a two-shift period. The physical link between

> the operator and the equipment can be broken because both the part loading and unloading and the tool control become support functions, rather than on-line requirements. Through asynchronous scheduling and buffering, the timelink can also be broken: the operator does not have to be at the machine to participate in the operation. This uncoupling of physical presence and time presence can improve the effective utilization of the equipment.[12]

As a result, the systems are designed so that "the human function is to load castings (or forgings, etc.) into one end of the system and remove them at the other," according to Moshe Barash, a leading FMS researcher at Purdue University.[13]

In many cases, some spectacular productivity claims are made on behalf of FMS. The Yamazaki Machinery Corporation, the Japanese machine tool builder, cites some impressive figures for its flexible machine shop in Nagoya, Japan. The $18 million system has 18 machine tools, occupies 30,000 square feet of space, has a staff of 12, and can turn out 74 different products in 1,200 variations. A comparable manual system, according to the company, would need 68 machines, 215 employees, and 103,000 square feet to do the same job.[14] Messerschitt points to some equally striking savings with its FMS compared to NC machines that are not integrated into a system: 44 percent fewer machines, 39 percent less floor space, 26 percent less lead time, and 44 percent fewer workers for a total savings in annual costs of 24 percent.[15]

There is little question that significant productivity increases are possible but what these figures leave unstated is the possibility of some sizeable debacles. Glossed over or totally ignored in many of the productivity comparisons is the downtime of the FMS. In other words, the FMS may be far more productive when it's running, but if it isn't running half the time, these gains can quickly be lost. And this leads to another question: Would designs that incorporate more worker responsibility retard the efficiency of the system or would they enhance it?

The goal of eliminating all worker input results in systems of unbelievable complexity that are prone to breakdown and thus require considerable amounts of human ingenuity to keep running. Paradoxically, then, the drive to get rid of all worker input winds up requiring more worker participation. And, systems that strive toward fully automatic operation can create a pervasive feeling of alienation among the workers who are left, itself a critical shortcoming at those times when input is in fact needed.

Donald Gerwin and Melvin Blumberg have noted how the drive toward increased authority can result in a very costly and very intricate system.

> One aim of this revolution is to introduce a degree of control comparable to that in mass and process industries. Consistent with this aim, the trend in computerization has been toward more complex, large scale, centralized, integrated, and capital intensive production equipment.[16]

TRW's Borzcik concurs with this and warns of some of the problems of totally separating the operator from the machine:

> It adds substantial complexity to the system because it requires many other complementary technologies to be in place. . . . It is important that management understand the impact of an FMS. This is not an off-the-shelf item and can be full of risk. There are severe reservations as to whether industry can effectively adopt such a complex multi-discipline system on the manufacturing floor without intermediate steps such as the manufacturing cell or flexible transfer systems being fully evaluated, debugged, and implemented. The number of disciplines

> required for successful application of an FMS is large and diverse; technology requires effective hardware, software, algorithms, management, and technical personnel.[17]

The elaborate and interrelated nature of these systems raise some troublesome problems for maintenance and downtime. Robert A. Berdine, an FMS specialist at Caterpillar Tractor, lays out some of these difficulties:

> If one wants to develop a sincere feeling of inadequacy, one should attempt an in depth analysis of a system of this complexity. One rapidly progresses from the tried and true conventional methods of cycle time, load/unload times and material handling analysis, to intuition and bad advice from onlookers, and into pure guesswork. The number of variables and possibilities are overwhelming. In short, more sophisticated planning and management tools are required.[18]

Blumberg and Gerwin concur about the maintenance difficulties:

> The complexity of the new technology is staggering (even to one of the authors, who is a graduate engineer). Yet, electricians are being told that since computers are electrical, they should be able to repair them after several weeks' schooling at the vendor's facility. Mechanics who have never had to deal with anything more complex than a gear box are suddenly expected to diagnose and repair a state-of-the-art, electro-hydraulic and mechanical device costing on the order of eight to ten million dollars and not completely understood by its designers.[19]

The result, according to Blumberg and Gerwin, can be high downtime.

> Breakdowns of computers, material handling systems, and machine tools are a major contributor to lower machine utilization than anticipated, and in some cases lower utilization than less complex equipment. One firm, for example, gets about 15 to 20 percent less uptime from its CIM [computer-integrated manufacturing] than from its stand-alone equipment, thus reducing management's control over production schedules and compliance with customer deadlines.[20]

These problems are underscored by two pioneering flexible manufacturing systems: one at Caterpillar Tractor in East Peoria, Illinois, and the other at Allis Chalmers in Milwaukee, Wisconsin. Both systems are aging—the one at Caterpillar was installed in 1974 and, at Allis Chalmers, in 1971—and far more sophisticated approaches are available today. But these systems remain good examples of lingering reliability problems because there has been more than ample time to debug them. The Caterpillar FMS, among the most heralded of the pioneering ventures in this field, produces transmission cases and covers for motor graders. It took Caterpillar four years from the date the order was placed in 1971 until even marginal operation was achieved. In the first few years of production, a variety of problems accounted for utilization rate of only 20 to 40 percent.[21] Even more recently, the engineer responsible for its operation complained that the system has some "nagging problems," but he nevertheless maintained it is considered successful and has been well accepted.[22] When I visited the plant in 1979, one of the two shuttle carts that transfer parts was not working and a number of machines were down.

The Allis Chalmers FMS produces cast-iron tractor parts. The system was designed to be fully automatic but had to be scaled back, after several years of operation, in order to meet quality control requirements. The computer now shuts down certain sections of the system after every sixth part to allow for a manual inspection. One operator tells why:

> We have to hold tolerances of a thousandth of an inch on a lot of these parts and that often is not very easy to do. It's a constant battle with the heat and humidity, which can cause things to go out of whack. And as machines get older, their bases settle, their heads shift, everything loosens up but those tolerances are still there.

While this intervention makes the operation more efficient, the worker gains significant new input. The operator now has an influence over quality through the accuracy of the inspections and a new influence over cost by the speed at which the inspections are carried out. Downtime, however, remains a problem. As late as 1982, it ranged between 50 to 80 percent on some of the key machines.[23]

The search for a "foolproof" system can lead to tighter shop discipline to compensate for the technical failures of the system. Moshe Barash states the dilemma:

> Whether it is economically feasible, even if technically possible, to foresee and prevent all types of human errors is questionable. What undoubtedly will be necessary is to introduce a much higher level of work discipline in operating these systems. What has been said in terms of weapons and counterweapons can also be said in terms of foolproofing and human errors, namely, that no matter how foolproof a machine is designed, there will always be a human who will make an error never anticipated by the designer.[24]

The human cost of freezing the worker out of the system is that most jobs on the FMS are robbed of autonomy and satisfaction. This is particularly ironic since these are two of the more important qualities supposed to result from computerization. Blumberg and Gerwin studied the work attitudes and working conditions in the Allis Chalmers flexible manufacturing system in May 1980. They did extensive interviews with eighteen of the twenty people working on the system. Their sample included two foremen, two machine repairmen, six operators, six loaders, and two tool setters. The questionnaire consisted of 177 questions, some of which were open-ended. They then compared these results with various normative samples, including one of 1,515 employed adults representative of all occupations and industries in the United States. Some of the conclusions are startling:

> Our findings on job satisfaction indicate that most FMS workers are dissatisfied with important aspects of their jobs. Three of the four occupations scored below the normative sample for every factor except overall financial reward. All of the job classifications ranked comfort and resource adequacy below the norm. Three of the four occupations ranked challenge and chance for promotion relatively low indicating they feel they are not exercising valued skills and they view their jobs as dead-ends. Repairmen scored higher than the norm on a majority of the dimensions. Tool setters, operators, and loaders were dissatisfied with practically every aspect of their jobs. Operators scored particularly low on promotions, and loaders on challenge.[25]

On the question of autonomy—the extent the job allows choice in selecting procedures—all of the job groups scored below the normative sample. Moreover, work on the system is more stressful than the national average, with the most critical factor being an inability to use valued skills. With the exception of two operators, all the workers interviewed felt that they had skills they would like to use but couldn't. Finally, fourteen of the eighteen people thought health and safety was a problem, although ten believed it was not great.[26]

Al Adams has worked on the FMS at Allis Chalmers since February 1973. When I spoke to him in late 1982, he was a tool setter. He described some of the sources of frustration and underutilization that the operators feel:

> Basically, what they do when you start is they show you the difference between a red light and a green light, they tell you how to start and stop the machine, and that's all they want you to do. Now for the operator it's pretty monotonous work. There's a light on every machine and if something goes wrong, the light goes on. So the operator is supposed to just walk up and down and wait for the red light to go on. Yet, even when you're not doing anything they don't want you to sit down or look at a magazine or something to break the boredom. All they want you to do is pace up and down.

The severing of the operator from the active control of the process can lead to emotional distress and an increasing lack of concern over what is taking place in the system. According to Adams:

> The work gets on your nerves. You become apathetic after a while so you don't want to do anything even when something goes wrong. Since you have no direct involvement with the workpiece, you don't take any personal pride in what happens to it. It's even unloaded in a separate area by somebody else. So guys stop caring if the part is good or bad or whatever. There's got to be a better way of keeping a guy occupied and giving him a feeling that he's doing something.

Although theoretically designed out of the system, when things do go wrong, substantial operator input may be needed.

Some FMS operators have shown unusual curiosity to learn more about the system and a real competence in developing unofficial methods that clearly improve operations. Ironically, these productivity enhancing methods come in spite of machine design and management desires. Adams tells of some of his early experiences:

> When I first started on the system I was working third shift. Things were pretty loose and there was a cooperative engineer, so I learned all I could about the FMS. I learned the assembly language and how to do NC programming. I did it because you want to take pride in your work, you want to know what you're doing and be able to fix something. But when I went on first shift, which I wanted to do because of the hours, the learning process came to an abrupt halt. They don't want you to do anything like that.

Nonetheless, Adams and his coworkers wind up intervening at key points to keep operations going smoothly. When a part goes out of tolerance, for example, the operator may call the part program up on a CRT and make the necessary alterations to get production running again. Adams has gone considerably further than this and developed a number of computerized systems to make his job more efficient. One system tracks the tools he orders and cuts the ordering time at least in half. Another system enables him to more effectively sort and deploy cutting tools on the shop floor.

Some of these problems could no doubt be remedied by a better organization of work without any major technological change. But, there are also problems that are rooted in the design of the system itself. In the pursuit of super automation, this FMS and others like it are designed with little concern for the attitudes, skills, and needs of the workers involved, particularly if these needs conflict with managerial authority.

Blumberg and Gerwin agree that even from a managerial view a sharp change of direction is necessary:

> What are the implications of our findings for the automatic factory concept? Our data indicates that such facilities are still a long way from becoming a practical reality. Until equipment can be brought on line which is completely reliable, which can procure its own raw materials and dispose of its finished products, the behavior of humans will continue to be important. System designers, in their haste to develop smoothly functioning systems free of human variability have committed the tragic flaw of overlooking the simple fact that humans are also critically important sources of control for system variability. It is time to recognize this and give greater consideration to the design of transitional systems where humans and machines interact rather than to the design of increasingly complex, less reliable technology where humans merely cope.[27]

Although the concept of flexible manufacturing has created a lot of excitement in the metalworking community, it has had little direct impact so far. There are probably fewer than twelve of the most advanced systems in the U.S. today. Currently, FMS accounts for only about 1 percent of NC production. Nonetheless, the concept has had a significant indirect effect. As *Iron Age* put it:

> Despite the fact that fewer than 25 true flexible manufacturing systems are operating in the world today, despite the fact that most of those systems have never realized the potential predicted for them, despite the fact that no one really knows for sure just what an FMS is yet, talk of flexibility can be heard everywhere.[28]

All the talk notwithstanding, the reality of the automatic factory has so far proven more difficult than the dream. As the reliability of these systems improves, however, the future could be different, but for the workers involved, not necessarily any better.

ROBOTS

When *Time* magazine gave an award for "Man of the Year" to the computer in 1982, the robot surely must have been the runner-up. Its homely appearance—compared more often to a metallic ostrich than a sleek computer-age automaton—had already burst forth on the cover not only of *Time*, but of *Business Week, U.S. News and World Report, Fortune*, and several other periodicals as well as in countless newspapers and on TV shows. The robot has become the symbol of new forms of automation in the factory, dramatizing the versatility, scope, and power of computer technology. At the same time, the robot has also raised serious questions about the consequences of technological change, particularly in the area of employment.

Robots have achieved all this attention at a time when U.S. manufacturing is in serious crisis. As a result, many observers have come to view robots as a panacea to improve competitiveness and restore high levels of profitability. Conversely, the failure of many industries is attributed to a technological lag rather than to any more fundamental shortcomings. There are many causes for lagging productivity, however, other than insufficient numbers of robots. As *American Machinist* put it:

> To make a direct connection between number of robots installed and productivity rates would be a gross oversimplification, of course. But such implied cause-and-effect relationships are probably indicative of the

finger-pointing that is creeping into some industries which—like automotive—are feeling increased pressure from overseas competitors.[29]

The undeniable triumphs of Japanese firms in the marketplace are often linked to their success with robotics, spawning endless comparisons between the number of robots on the job in the U.S. and in Japan. Unfortunately, most of these comparisons have been misleading at best and totally outrageous at worst. The Japan Industrial Robot Association maintains that, in 1979, 69 percent of the world's advanced robots were in Japan versus 16 percent in the United States. This translates into 14,000 units in Japan and 3,255 in the United States.[30] Although the percentage differences appear impressive, numbers this small could at best exert a very minimal impact on productivity. Some industry figures dispute even this lead by the Japanese. One of them is Joseph Engelberger, the former president of Unimation, the largest manufacturer of robots in the United States and, through its licensee Kawasaki, a major producer in Japan. In the summer of 1981, Engelberger told a symposium sponsored by the Office of Technology Assessment in Washington that rough parity in the number of robots existed in the United States and Japan. Perhaps of more importance, while some Japanese companies have utilized relatively large numbers of robots, others in the same industry have fared even better with virtually no robots. In 1980, for example, Toyota had 420 robots while Honda had only 5.[31] Robots may be vital for competitive success in the future, but they hardly explain what has happened to date.

There are compelling pressures to inflate the number of robots in Japan and deflate the number in the United States: The wider the gap, the greater the urgency to install robots quickly and the fewer questions asked about social cost. Rather than devising ways to cope with the displacement effect of robots, the public discussion is shifted to catching up with the Japanese. Under these circumstances, it becomes increasingly difficult to develop any real alternatives for a socially responsible use of the technology, thus potentially aggravating the pain of technological change.

Although the public discussion may be muted, the fact that robots are so clearly a replacement for human labor touches a real nerve with many workers. The question is job loss. Robots, of course, are only a subset of the computerized factory, one of many automated technologies that are being introduced. But unlike other technologies, say NC machine tools, which increase the productivity of a worker, the robot actually *replaces* the worker. The concern is heightened by the enormous range of application, spanning jobs from welders of car bodies to assemblers of typewriters.

Until recently, the robot industry has grown at a glacial pace. Over twenty years of development and commercial marketing have resulted in only 8,000 or so robots in the field by 1982. Today, however, the growth curve is bending upwards in an exponential fashion. Thomas O. Mathues, vice president of manufacturing development at General Motors, maintains that a "critical mass of driving forces" in terms of reliability, sophistication, and cost are now propelling the industry's development.[32] The result is that many analysts feel robots could expand from a $150 million-a-year market in 1981 to a $2 billion-a-year market by 1990. If peripheral equipment and applications are also included, the market could be $4 billion-a-year.[33]

This section will look at where the industry is going and the consequences of its current path of development. It will focus on the state of the technology today, the factors guiding its introduction and use, and on some of the likely impacts on both employment and the structure of work.

To begin with a frequently asked question: What is the definition of a robot? Mechanically, a robot is hardly a revolutionary departure from conventional forms of machine-tool technology, although the mechanics are becoming increasingly sophisticated. It is basically a multi-jointed arm capable of moving a payload to a specified location. What is different about robots, however, is that the location of the arm can be varied by providing the machine with new instructions. In other words, the robot is a general-purpose machine that can be programmed to do a wide range of specific jobs.

The flexibility is extraordinary. The same basic machine can be adapted to load and unload machine tools, reach into a heat treating furnace, assemble an automobile water pump, spray paint an appliance, weld a truck body, and even inspect a complex manufacturing operation. At the General Motors J car assembly plants, for example, robots use laser range finders to check the final dimensions on car bodies. The flexibility also means that even a major change in the product can be accommodated with minimum changes to the production technology.

At the Lakeville, Georgia, assembly plant, for example, the same robots went from building mid-sized Chevrolets in one model run to welding small Chevettes in the next. In contrast, much of the machinery would have had to be reworked or scrapped with conventional automation. As a result, a robot changeover can take one third of the time or less than it takes to convert fixed automation. In some applications, a robot can be transferred to an entirely different operation. If the welding line is shut down, the idle robots can be used to load machines or do assembly work with few modifications.

Although very versatile, the first generation of robots are unable to respond to their environment. As *Machine Design* magazine put it, "They are blind, deaf, dumb, and limited to a few preprogrammed motions. But in many production line jobs, that's all that's needed."[34] While this doesn't say much for the design of most production jobs, there are nonetheless some serious limits for robots that can't "see" or "feel." Once instructed to perform a given job, they will keep on doing it even if something around the robot isn't right. If a part is misaligned, for example, the robot might pick up "air," or crush the part, or perhaps smash something else. Moreover, seemingly simple tasks for a person, such as grabbing a part out of a bin, are impossible for a robot without some sensory feedback. Even with sight, however, a sense of touch is necessary for many assembly tasks. Otherwise, there is no way of telling if two parts are barely touching or have thousands of pounds of force between them. In addition, trying to fit a peg into a close fitting hole requires some feedback to avoid jamming the peg or gouging the hole.

To fill these technical gaps, the development of "smart" robots is proceeding at an intense pace. GM, for example, already has a simple vision system called Consight in operation at a number of plants. The system does material handling, sorting, inspection, and basic part recognition. In a dusty Saginaw, Michigan, foundry calling forth visions of nineteenth-century ironworking, a Consight system uses a camera, a computer, and special background lighting to sort as many as 240 different castings at a rate of 10,000 parts per hour. Systems such as Consight, however, see the world in two dimensions and recognize the part by its silhouette. This is fine if the part contrasts with its background, but in the real world this doesn't happen very often. Research, therefore, is now being directed toward systems that see parts three dimensionally in a way that is similar to human vision.

The state-of-the-art in tactile sensing is also moving ahead rapidly. Marc Raibert, a researcher at Carnegie-Mellon University, maintains that "if the reliability problems of tactile sensors were solved, perhaps 75 percent of robot arms would use them."[35] In the laboratory, "artificial skins" already can distinguish between the "feel" of different parts and have comparable resolution to a human finger tip. At an MIT laboratory, for example, a robot finger is able to tell the difference between six types of parts including screws, washers, and cotter pins.

While the individual robot is capable of increasingly spectacular feats, the full power of the technology is only realized when the robot is tied into larger manufacturing networks. Since the robot relies on miniaturized solid state electronics and digital logic, the basis is there for communication links that pull together into centrally directed systems robots, NC machine tools, conveyors, computers, advance sensors, and other technologies. The technical problems in carrying this out, however, are immense. Nonetheless, the movement in this direction was underscored by the GM manufacturing staff in a March 1982 letter to 200 robot builders and vendors. The automaker called for robot designs compatible with its computer networking system even though the final plans for the systems had yet to be completed.[36] Already the cost of the robot itself frequently accounts for only 30–50 percent of the total expense of an application. The rest of the cost comes from auxiliary machines and computer systems as well as engineering time.

One of the most advanced robotic systems in use today is a welding installation called robogate. Designed by Fiat, the system is used by Chrysler to weld the bodies for its K cars. Sixty-eight welding robots and other forms of fixed automation perform 98 percent of 3,000 body welds automatically. Ironically, the state-of-the-art system is located in Chrysler's aging Jefferson Avenue plant, built in 1907 by the Maxwell Motor Car Co., and located on the depressed east side of Detroit. Inside the deteriorating structure, a visitor is confronted with a maze of overhead conveyors, automatic parts shuttling devices, storage areas, computer consoles, and robots. The only thing that seems to be missing is people.

At the beginning of the final welding line, a few workers

are present who toy tab together (basically folding sheet metal protrusions into slots) the three basic body components—the two sides and the underbody. Then a conveyor shuttles this jury-rigged assembly to the heart of the system, a framing fixture called the robogate. Here two sturdy metal frames cradle the assembly, holding the parts in exact position while eighteen robots weld the body together. After this, more welds are added by thirty-two additional robots on the respot line. On each side of the car, sixteen mechanical arms appear to be moving through an awkward ballet, showering sparks after a quick jabbing motion to the car body. All the robots are electronically monitored. If there is a missed weld or a breakdown of some sort, a central computer signals an operator so that the problem can be located and the missing welds performed manually at the end of the line.

Mini-robogates feed the main welding line. The body side assembly is moved between various work stations by a monotractor, a carrier with an independent, signal controlled motor that is programmed to stop at the appropriate stations. After a body side is completed, it is automatically deposited into a computerized storage system designed to give the operation additional flexibility. "It's a manufacturing rubber band that provides a little give if something else goes wrong," according to Richard Vining, a former Chrysler vice-president for manufacturing who was instrumental in setting up the system. The storage area can absorb excess production from the subassembly areas or it can supply body sides when the subassembly areas are down. In addition, there is a computer-controlled storage area for the bodies themselves so that a breakdown in the robogate does not paralyze the rest of the assembly plant.

Some industry figures view the robot as only one particular type of end terminal for a computerized system. The central thrust of their activities is developing the system rather than the robot. Philippe Villers, president of Automatix, a small Burlington, Massachusetts, robotics firm, has been instrumental in developing this view.

> We believe that the tying in of so-called intelligent robots with smart microcomputers and with smart sensors such as vision means that we're really in the computerized systems business. Some of the terminals happen to be robots.[37]

This emphasis on robots as a subset of computer technology is reflected by the history and structure of Automatix. It was formed in 1980 by nine people, six of whom had substantial experience with other high technology companies, and three of whom are experts in robotics. Villers, himself, was a co-founder and senior vice president of Computervision, the largest CAD company in the United States and a firm that has gone through an extraordinary growth curve since its founding in 1969. Automatix's systems approach has attracted some impressive venture capital support, including $5.75 million from a group made up of, among others, Harvard and MIT. Ironically, although Automatix is considered a robotics firm, it manufactures none of its own robots, buying these from other firms and concentrating instead on developing sensors and software.

The central motivation for installing robots—either singly or in systems—is economics. According to Joseph Engelberger, "The basic production problem to which robots are addressed is the reduction of cost by eliminating human labor."[38] And robots are becoming cheaper than workers in a widening array of occupations. At General Motors, for example, wages rose 240 percent between 1970 and 1980 while the cost of purchasing robots increased by only 40 percent. In 1980, the total cost to an automaker of buying and operating a robot on a two shift basis for eight years was about $6 an hour while the total compensation cost for an auto-worker was $20 an hour. Furthermore, GM predicts that the annual cost of robots will rise 3 percent a year compared to 9 percent a year for labor in this decade.[39]

A number of factors serve to moderate the rise in robot costs. As more firms enter the market, fierce competition will likely hold prices down. Some of the newer and more aggressive manufacturers undoubtedly will employ learning curve pricing, a strategy that seeks to create a larger market by lowering the price of robots. This then provides the economies of scale to reduce manufacturing costs. As volume increases, mass production techniques are certainly feasible, since the most sophisticated robot is less sophisticated than an automobile. Moreover, stable prices for increasingly sophisticated electronic controls help keep the overall price of the units down.

Robots also have some impressive advantages vis-à-vis fixed automation. Since a robot is a general-purpose machine, large amounts of capital can be spent debugging and improving the basic design. This highly developed model is then adapted to specific applications rather than custom designing a machine for each new use, which is often the case with fixed automation. The basic robot is also instantly available as a building block for a more advanced system. Rather than having to design special-purpose machines with all the lead time that this requires, robots can be adapted to highly specific functions through software and a minimum of tooling. The same robot, for example, can be used to weld cars or unload machine tools.

Another oft-cited contribution is higher quality. One consequence of the stunning repeatability of this technology is the removal of some previous product design limitations. On an auto body, for example, redundant welds are specified to compensate for any a welder might miss or do below specification. With robots, this becomes unnecessary.

Robots themselves are becoming increasingly reliable and

long-lived. On many current applications, the robot is ready for operation 98 percent of the time or better. (The pacing factor, however, has been the reliability of the peripheral equipment.) The average productive life of a robot is now estimated at between 40,000 to 60,000 hours. Under the right circumstances, even this has been exceeded. At the GM Lordstown Assembly Plant, for example, the welding robots have accumulated over 80,000 hours on the job. In fact, while the second robot ever built by Unimation is in the Smithsonian, the seventh is still running.

In view of all these managerial benefits, there is little question that the industry is poised for explosive growth. In the process, its character is being transformed. In the past, robotics was dominated by relatively small firms whose basic expertise was in metalworking. In 1981, for example, the largest robot producer, Unimation, was a subsidiary of a small conglomerate called Condec. The second-largest robot producer was a division of Cincinnati Milacron, the largest machine tool builder in the United States, but a medium-sized firm at best by the standards of the rest of manufacturing.

Now a new breed of powerful, highly aggressive, and technically advanced firms are entering the market in record numbers. Some of these firms are highly integrated manufacturing corporations with enormous resources, such as General Electric, Westinghouse, Bendix, and United Technologies. Others are major robot users from the automobile industry such as VW, Renault, and General Motors. And finally, some of the new competitors, who have either entered the field or are near to making the leap, are world leaders in computer or semiconductor production such as IBM, Texas Instruments, and Digital Equipment Corporation.

When major robot users also become robot manufacturers a potent synergy is created that could speed the diffusion of the technology. This market trend also underscores the highly integrated nature of all computerized technologies in the workplace. As we have seen, General Electric, the largest user of CAD/CAM equipment, is now seeking to become the largest producer. The company spent over $500 million in 1981 alone buying up and developing companies that cover every sector of factory automation from Calma, a large producer of CAD equipment, to Intersil, a manufacturer of microchips. A central part of GE's corporate strategy is entering the field of robotics. GE's haste to become a major force has led it to license advanced robot technology from existing manufacturers worldwide including Japanese, German, and Italian companies. Based on this approach, GE was able to present eleven different robot models, all based on imported technology, at the Robot VI exhibition in Detroit in March 1982, one year after a press conference announcing its intention to enter the field.

Even the mammoth General Motors corporation is following a joint venture strategy. In March 1982, it announced an agreement to enter the robot business with Fujitsu-Fanuc Ltd., a major Japanese producer of robots and other automated equipment. GM projects the venture will have sales of $50 million annually as early as 1985.[40] This represents a sharp departure from past practice in which GM has always made purchases from robot manufacturers even if it had a major role in the developmental work. This new strategy stems, in part, from a desire to develop needed expertise and, in part, from a need to establish a benchmark to judge other manufacturers in an area that is pivotal to the corporation's automation plans. For these reasons, GM purchased a 10% stake in Automatix for $12 million in August 1984 with an option to acquire a further 10% in the next several years. Westinghouse has chosen to increase its muscle in robotics by purchasing Unimation from Condec for over $100 million.

Giant corporations such as GM and GE are also using real muscle to diffuse robots internally. Large manufacturing concerns frequently give their divisions considerable autonomy as to what type of technology to purchase. Therefore, it takes more than a simple edict to get robots into the factory. At the General Motors Technical Center in Warren, Michigan, for example, there is a robot laboratory that is used as a demonstration site for a variety of new technologies. In 1981, 6,000 GM engineers and managers from around the world were brought through to see demonstrations of thirty-five different types of robots and to consult with the corporation's experts. Since 1980, GM has also had a Corporate Robotics Council to coordinate its many activities in this area. General Electric has followed a similar approach with a centralized laboratory and a powerful corporate-wide coordinating body.

As the development of robotics becomes increasingly controlled by large multinational companies, many of the smaller innovative firms that mushroomed up in the industry may lack the resources to survive. This means that the technology could become increasingly molded to the purposes and needs of the large firms themselves at the expense of the requirements of the small- and medium-sized firms in manufacturing. The alternative ways of using robots in the workplace could become more limited as a result.

While the growth possibilities appear very bright, there are nonetheless some factors that act as a brake. The limited capability of the first generation of robots has meant that the technology is often difficult to install onto existing processes that were designed for humans. But there are three other important constraints: the experience necessary to install and maintain the technology, capital formation, and resistance in the workplace.

A firm must go through a critical learning curve to use robots effectively, particularly in a complex system. Among other things, it requires technical expertise and experience, which can be scarce commodities in a new and rapidly growing industry. Chrysler, for example, did not leap from conventional technology to its highly automated robogate system in Detroit. Robogate was preceded by a number of increasingly complex approaches in which Chrysler gained both an experience base and needed know-how. In 1975, the corporation began using robots for respot welding in its Delaware plant, a relatively simple operation. In the winter of that year, Chrysler manufacturing managers decided to introduce a more complex systems approach—an early robogate—at the Belvidere, Illinois, assembly plant to produce the Omni and Horizon, new front-wheel drive subcompacts. Finally, after all of this, the decision was made in early 1977 to install an advanced robogate system in Detroit to be ready for production on the new K-body car in the fall of 1980.

Even if the experience is available, the capital might not be. A robot represents a capital expenditure that is easily postponed, especially since most jobs that are done by a robot can either be accomplished by a worker or by other mechanical means. If a company needs a new lathe, however, there are few other alternatives. According to Engelberger, "Robots are innovation by invasion. Since the job can be done in other ways, the price has to be right for robots to be used."

Finally, resistance in the workplace could slow down the introduction of robots. Resistance from middle management has already been a troubling concern of robot proponents. Production supervisors, for example, are often less than enthusiastic about unproven technologies when they are pressured to meet tough manufacturing targets. And while labor has generally been very cooperative in the introduction of new technologies—some might say too cooperative—there is always the specter of resistance.

All these constraints notwithstanding, the potentially explosive growth of the industry raises some critical questions about the ways in which workers will be affected. The leading concern is job loss. On one level, the overall impact of robotics does not appear terribly alarming when it is spread across the entire economy. If the market is $2 billion a year by 1990, for example, and the average robot costs $50,000, this would mean annual sales of about 40,000 units, hardly a disconcerting figure in the context of 20 million or so manufacturing jobs. On another level, however, robots are only one of a wide range of powerful labor-displacing technologies that will be affecting the workplace. Robots' impact on employment must be considered along with numerical control, word processors, scanners in supermarkets, automated warehouses, computer-aided design, and a host of other advanced technologies. Moreover, the impact of robots themselves will be concentrated in a few key industries such as auto, which in turn are located in a few midwestern states. Today, over 50 percent of all robots worldwide are in the auto industry, and in the United States 80 percent of auto employment is located in five states. As Stanley Polcyn, a vice president of Unimation, puts it: "The real social problem is that we're looking to wipe out the unskilled jobs, and, if we don't recognize that fact, we're aiming for severe social upheaval."[41]

Rather than trying to assess the overall impact of robots on employment, an analysis dependent on a large number of uncertain variables, a sharper indication of the technology's impact can be found by focusing on the strategic plans of a few key companies. General Motors is an excellent place to start. The first application of robots in an assembly plant was at Fisher Body in Norwood, Ohio, in 1967, and in the next thirteen years fewer than 300 robots were installed. In 1981, however, in the midst of the worst crisis in auto since the 1930s, GM announced that it planned to buy 20,000 robots in the next decade. At the beginning of the 1980s, the average robot in a GM assembly plant displaced 1.7 workers; and in a three-shift manufacturing plant, it displaced 2.7 workers. These figures are on a net basis, that is, they include all the additional workers that are necessary to install and maintain the robots.[42] This means that over 40,000 workers could be displaced at GM by this technology alone. Put another way, robots could eliminate the equivalent of Chrysler Corporation's total hourly work force in the United States in 1981.

This potential displacement is sobering, but these figures may in fact understate the case. For one thing, these displacement calculations are based on GM's experience at the beginning of the 1980s. Robots that come on line in five or ten years, many with vision and tactile sense, will undoubtedly be more productive than their first-generation counterparts. And, as GM gains more experience with the technology, fewer maintenance and support staff may be required. Also, as the auto industry begins to recover from the debacle of the early 1980s, the corporation might choose to purchase more robots. In fact, Roger Smith, the chairman of the board at General Motors, has stated that for every dollar-an-hour rise in the wages of a UAW worker, 1,000 more robots become economically feasible.[43]

General Electric also has ambitious plans. In the June 1981 issue of *Business Week*, GE executives made the startling prediction that the robotic technology existed to eliminate half of the 37,000 jobs in the company's appliance division. GE has since publicly revised this estimate, claiming that while it may be technically possible to eliminate this many jobs, it was economically not feasible. This reversal came in response to the furor the original prediction created. But some internal studies reportedly indicate that the displacement threat remains very real. One participant at a forum on robots sponsored by the Office of Technology Assessment reported that a GE internal study indicated that of $900 million a year in assembly done in a number of divisions, one third could be automated today, and two thirds by 1990.

Viewed in the context of the larger economy, the displacement due to robotics is particularly troublesome because of the lack of employment alternatives for the workers whose jobs are eliminated. Even if the economy does well in the aggregate, thereby generating enough jobs to avoid large-scale unemployment, there are no bridges between areas of despair and regions of opportunity. In other words, unemployed autoworkers in Detroit do not benefit if there is a shortage of computer programmers in Boston. For the individual, retraining and relocation allowances for this kind of a leap hardly exist. Moreover, what happens to the communities whose employment base becomes permanently eroded? While some individuals will no doubt be able to find other options, those with the fewest alternatives will become the core of what's left of the community—a social disaster area in a decaying industrial infrastructure. Without broader measures, displaced autoworkers will more likely find themselves at lower paying jobs in the service sector—operating microprocessor-based cash registers at McDonalds rather than writing software at Wang.

One strategy to moderate technological change, often proposed to labor, is holding down wages. This may defer the introduction of robots somewhat but hardly solves the overall employment problems. Moreover, as robot technology develops, wages would have to be so low to be competitive that demand in the economy could be seriously impaired, let alone the living standards of the workers involved. One MIT research group, for example, modeled the cost of some automated systems against wages for a midwest manufacturer with a total wage cost of $5 an hour in 1980. Although the group concluded that automation was not cost effective under these circumstances at the time, the day was not far off when it would be.

While many are concerned about the displacement robots could cause, the technology is often portrayed as an unalloyed bonanza for improving the quality of life on the job. Robots are presented as eliminating the hazardous, unhealthy, and soul-deadening jobs in the factory. The first generation of robots, in fact, often did just that. They were deployed in welding, paint spraying, and diecasting—all unhealthy and potentially hazardous. This deployment, however, was more a result of the limited capabilities of these robots and the cost effectiveness of these applications than any overriding humanitarian concerns. As Joseph Engelberger put it:

> When one talks about harsh environments it is easy to conjure up sociological benefit arguments. If a job includes a health hazard, perhaps the robot's cost is not critical. In the real world this is false reasoning. The standard for the cost of the robot has to be something less than whatever it is that will induce a human being to accept the working conditions. Barring actual legal action, manufacturers simply won't introduce automation that is not cost effective in comparison with human labor.[44]

With return on investment as the leading criteria, the second generation of robots will be targeted for some of the most desirable production jobs in the factory—light assembly, machine loading, and inspection. Less than one third of GM's projected robot purchases in the coming decade, for example, will be doing welding and painting. The rest will be doing what are considered the better jobs. Moreover, workers themselves have little say and less power in determining where robots will be installed.

Robots also make possible tighter control in the workplace. According to Engelberger, "rationalizing the workplace provides the greatest secondary benefit" after direct economic gains.[45] Since robots require an ordered environment, they provide the perfect impetus to reorganize all surrounding jobs in the factory as well. According to Kenichi Ohmae, a director of the consulting firm of McKinsey and Co., "There is no point in introducing robots unless your factory already has disciplined production floors and standardized work procedures." Ohmae maintains that the following questions must be answered affirmatively before robots can be used effectively:

> Are workers closely observing standard work procedures? Are the right materials delivered to the right place at the right time? Are quality control standards applied to all components and procedures, including accuracy and tolerance in component size? Are production machinery, jigs, and tools maintained properly? Do workers properly operate stand-alone machine centers, transfer presses and the like? Robots are only as effective as their working environment.[46]

As robots are integrated into systems, the system itself could be used to control the pace of work in the factory. *Production* magazine put it rather boldly: "Whenever you have a production machine that is 'operator paced' you have an opportunity for improvement."[47] Unimation made the same point in an ad that ran in many of the leading trade journals. The headline maintained that "UNIMATE ROBOTS GIVE YOU ALL THE MACHINE-TOOL PRODUCTIVITY YOU'VE PAID FOR BUT NEVER GET." The ad then stated that "fast cycle time on a machine tool doesn't give you all of its high production potential if the men at the machine don't load and unload consistently and fast."[48] Autoplace, another robotics firm, advertises its robots under a help-wanted headline:

> Versatile Worker to Improve Productivity and Increase Accuracy in Parts Handling Operations. Willing to work days and nights; no vacations or sick leave. Job environment may be tedious or hazardous. Must offer flexibility, dependability and have perfect vision and concentration. Production applications unlimited.[49]

The theme of control receives its ultimate expression in a prototype system for light assembly developed by the General Motors Corporation. The system, called PUMA (Programmable Universal Machine for Assembly), may be the most complete integration of worker and machine since the automatic lunch feeder in Charlie Chaplin's *Modern Times*. PUMA combines light-duty robots (also called PUMAs), parts feeders, transfer machines, and people. According to *American Machinist*, "GM thinking is that robots and people can work together on assembly jobs—and that they should be interchangeable."[50]

What does this interchangeability mean? For one thing the PUMA robot is designed to occupy the same space as a person. GM's promotional literature for the system shows a worker squeezed between two robots on an assembly line. The pace of the line is determined by the mechanical arms. The worker loses even the limited autonomy of working up or down the line that is present on a conventional system. The robots will be programmed to do their jobs by engineers off the floor, following the example set by numerical control, where machine tools are effectively controlled from the engineering office. The recorded program will then be transferred from the methods laboratory to the floor where authorized personnel can make minor corrections if necessary. (This is in contrast to many existing robots, which are programmed by an operator leading them through their paces.) If there is a malfunction of the robot, the system is designed so that the mechanical arm can be pulled off the line and a human worker "inserted" in its place. The human would then be doing a job designed and paced for a robot while the robot itself was being repaired.

Why is the person there at all? According to General Motors engineers, "It was decided to retain human beings in the proposed system until vision and tactile systems were perfected and became economically feasible."[51] While the perfection of these systems may be some years off, as we have seen, GM already has quite a number of vision systems already in factories. Although PUMA as a system has not been installed anywhere yet, it clearly reflects GM's thinking in this area.

The relation between robots and the work pace of humans also has been explored as part of a larger study on robots in industry carried out by a German research institute.

> The conclusion yielded by the study is that we cannot speak of a thorough improvement of the working conditions; on the contrary, there are even some cases which are to be characterized as deterioration. This occurs, above all, in cases where the robots have assumed the task of handling the tools and left the worker only with the task of handling the material, that is, feeding new material into the machine.[52]

The study went on to state:

> Especially conspicuous was that in many cases the factory control of the work-rhythm was increased and the workers' freedom of movement was thereby further restricted. Thus, the possibility for the worker to regulate his own work, to set a faster or slower pace—obviously within certain limits—was reduced.[53]

MANAGEMENT INFORMATION SYSTEMS

In late 1980, the Ford Motor Company began production in one of the most advanced manufacturing plants in the world. The complex is located thirty miles east of Cincinnati in largely rural Clermont County near the small town of Batavia. The main building of the mammoth $530 million facility, a squat windowless one-story structure, sits on a 350-acre tract of land near old State Route 32. Ford has named the road in front of the plant "Front Wheel Drive" and the address is "1981." The road's name comes from the new ATX automatic transaxle that is produced in the plant. It is a complex automatic transmission/differential combination for the front-wheel drive Ford Escort, a new generation of world car that Ford is marketing and manufacturing throughout the globe. The address is meant to honor the year the Escort was introduced. Inside, the plant is a marvel of mass production capable of spewing forth 500,000 transaxles a year using the latest in machine tools and material handling equipment.

Automation, of course, is nothing new for the auto industry. Machined parts have made their way down transfer lines for over thirty years and Henry Ford would have no trouble recognizing the drilling, reaming, milling, and grinding taking place. Although having the most advanced machinery is impressive, what really distinguishes the Batavia plant is the widespread use of computer technology to organize production. Interlinked computer systems create an electronic model of the factory in which management is provided with up-to-the-minute information about what is happening and a data base for longer term analysis and planning. When directives are issued, the way they are carried out can be instantly observed, evaluated from countless different angles, and the results made a matter of easily accessible computerized record.

How is the flow of information managed at Batavia? Almost all of the 740 machines in the plant are guided through their paces by programmable controllers (PC), a digital device with an easily changed stored program. Designed for harsh industrial environments, this cousin to the computer transmits data about machine operation to other controllers or to a central computer, forming the basic building block of an information system that spans all levels of manufacturing activity. The computer network, linked by "data highways," is made up of five separate systems: first, a machine monitoring and downtime network that tracks production flow and dispatches maintenance workers; second, a time and attendance system that monitors personnel data such as tardiness and absenteeism; third, a stock status system that locates purchased parts, raw stock, and tooling; fourth, an Automated Manufacturing and Planning Engineering system (AMPLE) that combines engineering information with production data; and finally, an energy management system that, among other things, is capable of lowering the heat and turning on the lights. Production management plugs into this information through seventy terminals scattered through the plant. These terminals—a TV-like screen and keyboard—are in the offices of each production foreman on the shop floor, in the offices of maintenance supervisors, in the six zone offices that are responsible for the major sections of the plant, in the toolroom, in the warehouse areas, and in key departments in the administration building.

The amount of information made available to management is staggering. AMPLE, for example, provides a data base for the entire manufacturing operation. It already includes entries for all 106 production processes in the plant, 22,000 items in the general warehouses and tool cribs, 4,000 gauges, and 3,000 manufacturing drawings. In the future, plans call for including critical information about machine operation and maintenance and, ultimately, for taking in personnel records, material control data, and financial information. Moreover, a system such as AMPLE provides managers and engineers with some unique capabilities for using this information. Production information is always current. If an engineer alters a production process at 11:00 AM, the change is available throughout the plant at 11:02 AM. In the past, these modifications were relayed through the slow and often unreliable transmission of memos and verbal commands. Further, production data is now centralized. In the past, it was scattered through various departments with, say, the most up-to-date information on a part in one area and the most recent plan to make the part somewhere else. Extensive cross-referencing makes all this data far easier to locate as well.[54]

AMPLE, of course, is only one system. Other less comprehensive approaches have been around the factory for some time. Manufacturing resource planning (MRP), for example, stems from a system introduced in the late 1960s by IBM for gauging when key materials are needed in production. Essentially, MRP seeks to work backward from the delivery date and the quantity needed of a certain product to determine the most effective scheduling for machine time, labor, and materials. Over 100 of these systems have been devised and they have been deployed in over 10,000 plants. One key benefit is the need for inventory goes down since parts can be brought to the production process as they are needed. In some plants, inventory reportedly has been slashed by over 30 percent and the cost of parts pared by 6 percent.[55]

Systems such as AMPLE or MRP require substantial feedback from the shop floor. But, information is not an abstract or a neutral quantity: What data is defined, how it is gathered, and the purposes for which it will be used are all socially charged questions. Information-gathering systems can be designed in a way that provides more data for autonomous and decentralized decision-making or they can seek to monitor every aspect of what a worker does on the job. The issue is not the value of timely information to coordinate production but the collection and use of data in a way that seeks to extend managerial authority. The latter approach weaves a net of electronic control around the worker, bringing the equivalent of assembly line discipline to those jobs that were technically impossible to police fully before.

Consider the design of machine and maintenance monitoring systems. A key motivation is increasing machine tool utilization. Downtime is viewed as a critical problem by Detroit and other mass production industries. If a transfer line is actually running over 60 percent of the time, it is considered a major success. The Ford Motor Company, for example, did a study of 154 transfer systems throughout its operations between 1974–80 and concluded that the systems were functioning only between 46–64 percent of the time. The downtime of all Ford operations was a rather high 23 percent. A variety of factors contribute to a machine not running—a lack of raw material, a broken tool, a malfunctioning machine, or an absent operator.[56]

One approach to eliminating downtime is to establish a much closer surveillance of the activities of the worker at the machine. According to Charles F. Carter, Jr., the manager of advanced machine-tool systems at Cincinnati Milacron, "As machines become more productive and downtime more expensive, all aspects of operator behavior, from attendance to job knowledge, will become more important."[57] But as manufacturing becomes more capital intensive, the percentage of workers not directly engaged in production rises. Indirect workers at GM accounted for an amazing 207,000 people or 42 percent of the work force in 1979. If anything, the drive is even stronger to extend managerial authority in this area. Two General Motors engineers in the manufacturing development program, Gary J. McCoy and S. M. MacMillan, maintain that the corporation "has done extensive developmental work in measuring and controlling indirect labor. Emphasis has been to provide better ongoing management controls for such jobs as janitorial, setup, cutter grind, oiling, maintenance, and most recently material handling."[58] And as McCoy and MacMillan put it, "If you don't measure it, you can't control it!"[59]

How do these systems work? At Batavia, a programmable controller links the machine on the floor to the central computer system through an electronic bridge that is called a "data highway." Every time the machine makes a part, or "cycles," it registers in the computer. Data in this form, however, would only swamp the system. "All of a sudden we know more about our business than we care to know," says Norm Hopwood, the manager of Computer Systems at Ford. So the system is set up to monitor exceptions to preset rules. When a machine doesn't produce a part within an allotted time, say two minutes, this fact becomes displayed on video screens throughout the plant, particularly in the foreman's office, and if the delay is serious enough it may be recorded on a computer printout at the end of the day. "If that machine hasn't turned over, we want to know what's wrong," maintains Hopwood. "Is the operator goofing off, is there a problem with stock, or is there a breakdown?" The foreman is dispatched to the scene to find out and the results are entered into the computer.

In the past, management was chiefly concerned with the output of a production machine at the end of the day. If a worker ran a machine that produced automobile axles, for example, the production quota might have been set at 200 axles per day. At the end of the shift, the worker was responsible for those axles and filled out a time ticket stating the completed production. If management felt that more could be produced, a time study expert was sent out to study how long the operator took on the job. Short of this—and the foreman watching the worker more closely—there wasn't much that could be done. And the foreman couldn't be everywhere at once.

The worker often devised ways to meet production in less than the allotted time. If so, the operator might press harder before lunch to produce 150 axles, and then have less pressure after lunch. Some supervisors became accustomed to delivering the extra parts to another section early and winked at them being produced in this way. In fact, if 100 axles were actually produced in the afternoon, it could ruin production schedules in other departments which often took account of the unwritten 150/50 output rate. Moreover, if something unexpected did happen such as a machine breakdown, the worker would have flexibility to work around it and still complete production. Within the context of a fixed production quota, the worker retained substantial control over the pace of the job.

Now the capability exists to analyze production operations as they are taking place with pinpoint accuracy. Data can be summarized, pulled apart, manipulated, and displayed in countless ways. The rate at which parts are produced is constantly monitored, for example, and one display can show the hourly production rate: "110 were made in the first hour, 101 the second hour, 109 in the third hour, et cetera." According to Hopwood:

> Without this system all we know is that a department can make maybe 600 parts per hour. With this approach we can also say that such and such is the bottleneck or that an individual machine is the bottleneck. Today, if we try to do that it gets lost in a shuffle of paper.

A monitoring system at Cadillac's new Livonia engine plant, in a suburb of Detroit, obtains about fifty pieces of information from each machine on a transfer line. Time is broken down into a variety of categories. The period a machine is waiting to make a part, for example, is not just "idle time" but time spent "waiting to load parts, waiting to unload parts, and waiting for the operator to push the cycle start button." It becomes possible to compare the length and nature of downtime between different departments or between different shifts. Moreover, the records of different operators can be compared with each other or reviewed for how they have developed over time. In addition to the time related data, the Cadillac system also records the number of parts completed, tool changes, rejects, inspection dimensions, and the capacity of automatic storage units.[60]

Monitoring systems are also finding increasing use in batch production. In an aircraft plant such as the sprawling McDonnell Douglas St. Louis plant, a single NC machine might be worth between $250,000 and $1,500,000. In the 750,000 square foot machine shop, one of the largest and most advanced of its kind in the world, a reporting system is linked to direct numerical control. The system gathers and transmits information concerning more than eighty-five machine tools "twenty-four hours a day, seven days a week." It divides machine tool activity into over 1,000 categories, including production bottlenecks and maintenance delays. A daily downtime report "lists all delays occurring by machine tool and shift during a reporting period." Special software in the system is set up "to report exception status conditions, such as excessive lunch, break time, and unaccountable hours." This data can become the basis for disciplinary activity.

> These reports allow an assistant foreman to evaluate performance by machine tool or machine tool family and to compare this performance with other work shifts within an area of responsibility. Problem areas are identified where corrective action is required. Daily and Weekly Summary Reports are also generated for review by Department Foremen and Machine Shop Superintendents.[61]

In the enormous Evandale jet engine factory, General Electric has introduced a similar system coupled to direct numerical control. GE engineers have catalogued fifty-one different reasons why an operator isn't running a machine and have incorporated the nine leading ones into the system. In particular, excessive setup time, use of the feedrate override, and being absent from the machine are monitored. Ultimately, the system may also track parts, attendance records, labor vouchers, inventory control, work schedules, and cost product history. Ralph Jansen, a manager at GE's Evandale jet engine plant, underscores an important reason for the systemization:

> We give the operator too many excuses for not doing his job. When an average operator runs into a problem, he doesn't want to use his initiative and ability to solve it. Let's say he can't find a tool for his machine in the crib, he'll spend a lot of time wandering around looking for one. That's why we want to monitor what's happening and systematize the operation.

At a Black and Decker plant in Tarboro, North Carolina, you don't have to be sitting in front of a cathode-ray tube to be in touch with the monitoring system. It is also hooked to red and green lights on workstation terminals. A red light flashes if the operator is away from the machine without authorization and other combinations of lights indicate the machine is running, the operator has requested assistance, or the equipment is not scheduled to run.[62]

The role of the foreman changes as well. On the one hand, the foreman doesn't have to leave the visual display unit. William T. Lesner, a senior process engineer at Cadillac, extols the virtues of the Livonia plant's monitoring system:

> Under today's circumstances, without the computer, how do you do your job? If you're like most supervisors, you end up spending a good portion of your time walking up and down the line trying to figure out the condition of your equipment . . . not really supervising.
>
> With the computer's help, however, you can immediately see the status of your entire line on the 19-inch color CRT setting on your back desk.[63]

On the CRT, a box represents every machine on the line. Lesner describes how the system works:

> If green . . . the machine is running.
> If green and blinking . . . it's running, but maintenance has a job open for this machine. . . .
> If yellow . . . it's in an idle or fault condition.
> If purple . . . the machine is in a manual or hand mode.
> If red . . . the power is off.
> Underneath each machine is critical performance related information that varies based on the status of the machine . . . in the case of a machine that is running, we show the number of pieces produced so far this shift and the machine's last cycle time.[64]

On the other hand, these systems often make it necessary for the foreman to be on the spot if something does go wrong. In a monitoring network at the Warner Gear plant in Muncie, Indiana, there is a terminal on every machine tool so that a worker can signal any problem without leaving. If a machine isn't running, the foreman must be there to insert a key that authorizes downtime. George Fenton, the Manager of Industrial Services at the plant, maintains that this feature is designed so that the supervisor is no longer obligated to "accept the operator's claims in order to maintain harmony in their mutual work environment."[65] This also insures that "the foreman must be on the floor rather than hiding in the office somewhere." If a disciplinary action results against a worker, it appears to flow from the system rather than being a subject of human choice.

Some supervisors are less than enthusiastic about the new work relations these systems bring. Their concern is that the information net used to highlight the workers' activities also displays what they themselves are up to. General Electric's Jansen states that "the foremen wonder how the system is going to be used against them. They feel they know what their operating problems are and if they can take care of them why should anyone else have to know?"

Some monitoring systems have met with considerable informal resistance on the shop floor, as we will see in more detail in the next section. Norm Hopwood points to the need for more sophisticated managerial methods in installing the systems:

> Four years ago, we installed a machine tool monitoring system to track the performance of about ninety machines in our Sharon, Ohio, transmission plant. It originally was installed in one machine zone as a pilot installation with expectations that it would eventually be expanded to the rest of the plant. It's still in that zone! Generally, the problems have not been technological; rather the problems have centered around people. We've since had successes in other plants, but the transmission plant project taught us a great lesson about the importance of the effect that people can have on the success or failure of a monitoring system.[66]

In one factory, workers quickly came to terms with a newly installed monitoring system. They devised a way to keep the machines running empty and recording. The worker would take a break and the machine would "cut air." For a while, everyone was happy: The workers could pace their job, and the computers recorded what management wanted to see. But then the number of parts recorded was compared with the number of parts actually produced, and the company countered by linking the computer directly to the machine motor. When a machine cuts metal, it draws more power than when it runs idle. Hence, management could tell when parts were being produced and when the machine was just "cutting air." No more unauthorized breaks. In another system, the computer is programmed to sense how fast parts are being produced. If a worker attempts to tamper with the machine, the computer will signal that the rate of production has changed and produce an alert on the TV monitor.

The web of control goes beyond the machine operator and the supervisor to include the skilled maintenance worker as well. In a large manufacturing plant, dozens of skilled trades may be necessary to keep production moving—electricians, pipe fitters, machine repair, toolmakers, millwrights, welders, and more. Monitoring systems can be designed to provide the repair person with valuable data of the history of a machine's breakdowns, in which case they monitor the machine, or they can be set up to deliver data on what the repair person is doing, in which case they oversee the worker. The financial stakes are quite high. Reducing the number of skilled workers at General Motors by 10 percent, for example, would have generated over $340 million in savings in 1980.

Skilled repair people are increasingly central to production for two reasons: The trend in machine design is toward more capital-intensive equipment to replace workers and toward more complex equipment to minimize the input of operators. Moreover, these heavy investments create strong financial incentives to run machines at full tilt as much of the time as possible, creating considerable strain on the equipment. When machines do break down, the pressure to get them back in service is enormous. Ironically, the drive to free production of the control of direct labor results in a new dependence on skilled trades.

The repair of industrial machinery has always been a varied and highly skilled task, relying heavily on the experience, insight, and judgment of workers. These qualities have made maintenance resistant to Taylorism. The job requires independence because it is the worker who has responsibility for these judgments. Moreover, maintenance workers have considerable on-the-job mobility, making close supervision difficult. In a large plant, for example, a breakdown may occur a half-mile from the nearest maintenance foreman, which means that the supervisor has to rely on the workers' own initiative and ability to get the job done.

Take machine repair. The most difficult part of the job is often diagnosing the problem. The first step is for the repair person to talk to the operator, the person most familiar with the day-to-day operation of the machine. If this doesn't pinpoint the problem, the next step is working through a routine inspection, all the while listening for any change in the machine sound or sensing any different vibrations in its cutting motion. A skilled worker will take all the time that is necessary on the diagnosis, knowing that once a machine is pulled apart, events can assume a life of their own. To a foreman, however, a machine repair person diagnosing a machine can be a painful sight because there is little sign of actual activity.

Skilled workers take a fierce pride in their ability to accurately troubleshoot and competently repair a complex breakdown. They are motivated to do a good job by the traditions of their craft and the respect of their peers. Moreover, the skilled worker may have to live with the machine. A shoddy repair usually means that the job will have to be redone often after the jury rigging has caused even more damage to the machine. This concern for quality, however, conflicts with the demands to push production through a plant as rapidly as possible, production management's most pressing goal.

Since they are taking responsibility for a job, tradespeople dislike any interference in determining how the job should be done. They strongly dislike time study and, in the auto industry, have had enough muscle on the job to prevent its use. If trades workers are harassed, their "judgment" can quickly disappear and they will only too gladly follow any instructions to the letter, often with disastrous results. This lesson was learned the hard way by a young supervisor on an engine block machining line. In order to speed up an especially troublesome job, he told a machine repair person to forget anything else and get the machine running. Feeling pressed by the foreman's instructions all day, the repair person was only too happy to oblige. After all, he hadn't been told to check the hydraulic lines. The one that was still unhooked bathed the foreman, the machine, and half the department in warm oil.

This independence conflicts with managerial authority. Consequently, computer technology is targeted to gain control in an area where the organization of work by itself has failed. "A considerable saving in repair time, and a reduction of training needs and (usually hard to get) skill levels, can be achieved if the maintenance operation is systematized and assisted by a set of computer programs," according to Alan Hills, the manager of the Ford Motor Company Productivity Office.[67] The challenge is how to systematize this enormously complex operation.

In a typical system, when a machine goes down, the foreman investigates the cause and then signals the maintenance office for the appropriate repair person, say, an electrician. The maintenance dispatcher assigns a priority to the job and requests a printout on the previous breakdowns the machine has had in order to aid the electrician in finding out what's wrong this time. The electrician is then sent to the machine. All aspects of the operation are quantified and visible on a video screen in the foreman's office: the time the last part was produced; when the foreman asked for assistance; the time the electrician was assigned to the job; how long the dispatcher estimates the job will take. On some systems, the worker "clocks in" when arriving at the machine and "clocks out" when leaving. After the repair is complete, the electrician enters into the computer what was done on the job. All this information is then summarized and recorded for later use.

A particularly sophisticated maintenance system accompanies the machine-monitoring setup in the Cadillac Livonia engine plant. Among other space age features, it has three color TV display units. The first shows the status of all critical conveyors, the second the current assignments of all available skilled workers, and the last the maintenance jobs that haven't been completed.

At the touch of a button the system is capable of presenting downtime from a variety of different angles. Consider, for example, the "Top 20" reports—four different views of the twenty most trouble-prone machines in the system. The first version pinpoints the machines that are down for the greatest number of hours, the second locates the most dramatic changes in performance—machines that slip from perfect operation to chronic problems, the third stresses machines with the most service calls, and the fourth emphasizes the machines with the greatest number of labor hours (this differs from the first version in that it stresses multiple trades on a single job).[68]

Skilled workers have expressed concern about the uses of all this information. While some data, such as the history of machine breakdowns, can be useful for doing a diagnosis, they note that all the data is passed through a managerially controlled network. The issue of power is underscored by the fact that in most cases hourly workers are prohibited from operating the dispatch terminals. The possibility is certainly there for a time study of the individual or a speedup of work in the entire department. Alan Hills admits that "the identification of a maintenance-technician goes into his data and records of his effectiveness can be retrieved."[69] One worker expressed concern that attention to quality on the job might conflict with the dispatcher's estimated completion time and thereby produce a poor record. Moreover, it becomes possible to compare the performance of one shift against another, or to play different plants off against each other.

These systems have sparked both formal and informal resistance. Bob King, president of UAW Local 600 and chairman of the UAW Region 1-A skilled trades council, has been particularly vocal against the use of these systems to do time study on workers. "In the old days, the Ford Motor Company had Harry Bennett's goons to police workers on the job. Now they are trying to use computers for the same purpose." In some plants, workers have refused to cooperate with anything they perceive as a potential time study. One skilled worker, when it came time to describe what he had done on a repair, consistently wrote in "trade secret." Others have written bogus or misleading information on their time tickets. In one Ford plant, workers were asked to wear beepers connected to the monitoring network so that they could be paged anywhere in the plant. Their response was that they would be unable to take responsibility for this "delicate"equipment in a hazardous industrial environment.

Nonetheless, there are claims of impressive productivity gains made on behalf of some systems. For one thing, managers are aware that there is resistance to these methods and strategies are devised to work around it. K. F. Noblitt, manager of Maintenance Services at Cummins Engine in Walesboro, Indiana, admits that "human error, omission and hostility are built into the data we collect." The solution is to slowly build up a valid data base that can then be used for comparative purposes. An early version of this type of system helped slash downtime at the Cummins plant, a nonunion facility. In 1974, 1,000 machines in the shop were down for a total of 100,000 hours. In 1978, downtime on 1,280 machines was cut to less than 50,000 hours.[70] At the GE Evandale plant, downtime was reduced from 11 percent to 7 percent.

The reasons for the increased productivity are more complex than may first appear. Does efficiency, for example, primarily occur because there is more information on the floor or as a result of management's tighter control? In some cases, gains may be achieved in spite of attempts to extend authority. Moreover, what is the long-term impact of trying to reduce the autonomy of jobs in which independence has been a source of creativity and initiative? Short-term gains could be followed by an erosion of skill and professionalism, a serious long-term cost.

TOTAL OPERATIONS PLANNING SYSTEM (TOPS)

The social issues involved in management information systems can be illustrated in the introduction of a computerized scheduling system for die construction in the mammoth tool and die building at the Ford Rouge complex. Introduced in the middle seventies, the Total Operations Planning System (TOPS) sought to provide managers with more information about what was happening on the production floor. This, however, was to be only the beginning. The system was also designed to bring more order and efficiency to operations and, therefore, has significant implications for the skills of workers and the distribution of authority in the factory. Not surprisingly, those whose skills and authority were being negatively affected didn't look upon the changes very favorably. In fact, a three-way struggle ensued between staff managers, production managers, and diemakers. Imagine a triangle. In one corner are the systems analysts from the staff of the Metal Stamping Division, the Ford group responsible for stamping plants and die rooms, of which the tool and die plant was a part. These staffers initially proposed TOPS and were responsible for implementing it. In another corner are the diemakers, who felt that their skills and authority would be eroded by the system. And in the final corner are the production managers in the plant who were perennially sparring with the diemakers but who felt that the system might undermine their own autonomy vis-à-vis their divisional superiors. All of this, combined with the intricate nature and technical shortcomings of the system itself, proved to be too much. TOPS was withdrawn nine months after it was implemented.

In a number of ways this case may be atypical: The product and the production process are unusually complex and the workers involved are exceptionally militant and sophisticated. Nonetheless, the ways computer systems are used to revamp relations of power and the importance of the shop floor response are both visible out of the wreckage. The peculiarities of this particular situation underscore the broader issues that are relevant to a much wider range of manufacturing operations. In exploring these issues, I will first review the diemaking operations, then examine the origins of TOPS, the technical details of the system, and what happened when it was implemented on the floor. I conclude with an analysis of a replay of TOPS that was introduced a year or so later and that sank even more quickly.

The production of dies is among the most complex of metalworking tasks. There are four or five basic kinds of dies and hundreds of variations. For a complex die, a huge casting, often weighing several thousand pounds, has to be laid out, machined, mated with hundreds of other parts and assemblies, and finally fitted into a high-tolerance unit that is capable of stamping hundreds of thousands if not millions of sheet metal parts. The logistics can be nightmarish for the six to seven hundred dies made each year at the Rouge die room. All the subassemblies have to be completed on time for the die to be ready on schedule and this requires substantial planning and coordination. The way the die is constructed depends heavily on diemaker leaders, union members who direct the activities of five or six other diemakers. As one manager put it, "Without their knowledge, there is nowhere you can look that will tell you how to build a die."

Although the tool and die building enjoyed a reputation as an efficient plant, the staff managers of the Metal Stamping Division had only the haziest idea of what went on while the die was actually being built. In fact, the further away a manager was from the shop floor the more out of control the entire process seemed. Part of this was due to the variability inherent in the production of complicated one-of-a-kind parts. The volume just wasn't there to fine tune production operations from the top since the next die would be slightly different from the one being worked on. A casting from a supplier, for example, might arrive with some extra metal requiring more machining than anticipated; or, when all the parts are assembled, additional grinding might be needed to get them to fit. Moreover, constantly shifting corporate priorities can throw operations out of whack. The decision to speed up the introduction of a new model means that its dies have to be rushed through production, disrupting the planned schedule. Furthermore, plant-level managers do not always issue accurate reports to the division. As one staff-level manager put it, "the foremen are always shuffling numbers around so they won't look bad on jobs where they're having trouble by charging the extra time to the jobs that are going well."

All of this in general and the reliance on the skills of the leaders in particular became a central concern of John O'Brien, a programmer analyst in the systems group of the Metal Stamping Division. O'Brien was the originator and a key player in the implementation of the TOPS system. He maintains the need for such a system occurred to him after a plant visit to work on a nonrelated labor system. He describes his impressions:

> While at the plant, I was talking to various supervisors and you get the normal feedback of, "my god, I need help with this or I need help with that, can you do anything?" Also it became apparent after talking to a number of people that the skill level of the workers that build the dies was decreasing. In the old days, a craftsman built a die from beginning to end and knew the entire process.

> Now they were becoming specialists in one area and none were familiar with the whole flow. We were also going through a cutback and union rules regulate that people move from one plant to the other within the complex based on seniority. So the leaders were being pulled out of the tool and die building.

Under the best of circumstances, the scheduling techniques disturbed O'Brien. In a manner reminiscent of Frederick W. Taylor, he felt that there must be one best way to make a die. Yet, if the plans for a die were given to the twelve superintendents and general foremen in the building, twelve different answers on how to build it would result. The influence of the leaders made things even worse. According to O'Brien "you talk to these people about scheduling methodologies and they have no idea what you're talking about." A further problem was that the construction of a die, which might take nine months to a year, was broken down into only four broad phases: doing the preliminary work, putting the shape on the die, assembling it, and trying it out. Within these four basic milestones, virtually everything else was controlled on the floor.

TOPS was designed to introduce more central direction into diemaking through computerized scheduling. To do this, however, the process of diemaking itself had to be more clearly defined and a more accurate method of monitoring the shop floor introduced. As O'Brien would later tell a meeting of diemaker leaders, "The idea is to take the top off the building, see where everything is routed, and then eliminate the bottlenecks by using a computer." The system had three parts to it: pre-construction planning, job completion tracking, and cost tracking.

The first phase was by far the most complex. It required an accurate, easily usable model of the diemaking process, simple in theory but a herculean task to develop. Once this was done, a code could be assigned to each operation, an optimum sequence for building the die selected, and the necessary time for each detail estimated. After this, the most efficient routing between the machines for the die components could be determined, and raw materials could be ordered with more certainty. Moreover, it would be possible to smooth out the work flow and improve the capacity planning for all the operations. The estimated time for each operation, for example, could be compared with the available machines and systems in the plant. Initially, these determinations would be based on rough estimates of previous experience, but the system would be refined with each die that was processed. A further selling point was that with everything so completely spelled out, virtually anyone could build a die, lessening the plant's reliance on highly skilled workers.

In the next phase, job completion tracking, the time each operation actually took and the progress of the die were reported back. Theoretically, any bottlenecks would be easily spotted and quickly corrected. In the last phase, the labor and material costs would be monitored so the actual performance of the plant could be compared with its projected performance. Ultimately, the goal was generating a code number that would enable the computer to determine automatically the best way to build the die and how long it should take. Additional plans called for tying this data base into a wider computerization of other production operations throughout the complex, such as the stamping plant, and throughout the Metal Stamping Division.

To implement a system as elaborate and potentially expensive as this, considerable managerial commitment was required. O'Brien made a series of presentations to the divisional general manager and other high-ranking managers, after which the basic idea was approved in late 1974. Then, the tool and die plant signed on. In the several years it took to develop and implement the plan fully, however, the plant management changed three times, and this was to have some serious consequences.

The complexity of trying to rationalize the diemaking operation proved to be staggering. During its development three programmer analysts worked on the project under O'Brien's direction. He is proud that everyone had at least a master's degree, "anywhere from mechanical engineering to a Wharton MBA." What they didn't have was any direct experience. To remedy this, a seasoned supervisor was recruited from the plant to work on the project, particularly to help develop a coding scheme for the dies. Even this proved inadequate, so during the 1976 strike between Ford and the UAW, the tool and die foremen, none of whom were on strike, were enlisted in the coding process. Many of the foremen were ex-diemakers and ex-leaders themselves and in any case had little to do during the dispute. Ralph Kuhn, the supervisor, had over forty years of experience in the diemaking trade. Although he worked hard and effectively on the project, he had serious doubts about its overall thrust from the beginning.

> From day one I felt it wouldn't work and it's simply because of the nature of the trade itself and of the people involved. If you could go ahead and have totally competent diemakers, all at the same level of competence, absolute die design that was perfect in every respect, castings that were perfect in every respect, everything flowing together, no mistakes by machine hands, no mistakes by the diemakers, no mistakes by anybody, maybe if you had total dedication, it would work. But that's utopia. It doesn't exist.

The day after the strike, a young diemaker was promoted to a salaried position to help coordinate the project. Stan Heide, a bright, personable individual, knew the diemakers and their habits well. He had served his apprenticeship in the plant and worked there since June 1969. His specific duties were to finalize the programs the foremen had written, help sell the system to the workers in the plant, and then coordinate its operation on the shop floor with the systems group.

Getting the knots out of the system, however, proved to be no small job. It took another three months of intensive work before the system finally was brought into the plant on February 14, 1977. The first try was to be a pilot program for five dies. Prior to the start date, a number of information meetings were called to explain the system to the diemaker leaders who would be instrumental in its effective use. The sessions were compulsory for all the sixty or so leaders in the plant. O'Brien dominated the first two or so classes, seeking to sell the plan as the latest in computerization, simply a method of enhancing efficiency without harming anyone. This seemed to fall flat, according to one member of the systems group, who stated that "the leaders tolerated the meetings because they could sit down for forty-five minutes and relax." After this, Heide tried a more down-to-earth approach in five more meetings on each shift, stressing the benefits to the leaders in carrying out their jobs. In fact, he scheduled some of these meetings with no other members of management present so the diemakers could be more frank.

The leaders expressed three strong apprehensions about the system: first, that it would erode their trade and eliminate the need for leaders entirely, second, that it would be used to monitor and time study all their activities, and, finally, that it would require a lot of effort to administer with very little payback for the leader. Moreover, some of the leaders wondered why if the project was to be such a boon to them they had been excluded from any of its initial planning.

The first fear was the most important. This was underscored by Jack Donovan, who had been a popular and well-respected leader in the building since 1965. At the time of the implementation, he was a bargaining committeeman and vice president of the unit and attended the orientation meetings in that capacity. According to Donovan:

> The main fear of the leaders was seeing their job going down the chute if the system were successful. They were going to take a readout sheet from the computer, and each operation of the die would have a code number on this sheet. All of this told you how to build the die in the proper sequence. You just go down the readout like A,B,C,D and you machine it and you mount the plates and mount the sections and you fit this and you do that, say, spot the pads. The end result would be when you build a similar die the next time, you won't need a leader.

Donovan expressed a widespread feeling by saying that the system would eventually make diemaking "more or less like following the instructions to assemble a swing set." Another leader complained that "they want to make us clerks instead of leaders." Their apprehensions were echoed by a third leader who felt that "if they ever develop a workable program and use it for five or ten years, we'll all wind up as robots." Ralph Kuhn understood and sympathized with these sensitivities:

> To standardize a craft and make it a semi-production operation is treading on dangerous territory. You're trying to reduce their pride and their own abilities. Reduce their status from the highest level of the factory down to a semi-production status.

Paradoxically, while threatened by the ultimate perfection of the system, some leaders were aggravated by its initial flaws. There was one common point of agreement with O'Brien: There are a lot of different approaches to diemaking.

But rather than viewing this as a liability, the leaders perceived it as an asset, given the variability associated with fabricating a complex one-of-a-kind part. As one leader put it:

> In diemaking, everyone thinks differently. It's easy to overlook these differences and just say on a program "mount the wear plates" but when and how this should be done can vary an awful lot depending on the die and whatever else is happening in the shop. With a formula, you would also miss all kinds of shortcuts.

Another leader complained:

> The operation numbers told you in no uncertain terms what order to follow. But, it was impossible to do things in that order so we skipped around. If they tell me how to build a die and I don't like it, I am still going to build it my way. Instead of doing number ten first, I might have to do fifty first and then go on to eleven, otherwise the die would never come together.

The second major objection centered on the time-reporting aspects of the plan. A leader used to fill out a time ticket to account for each worker's day, but time was allocated in a very general way. So for a particular diemaker, four hours might be charged to die A and four hours to die B. On the TOPS routing sheets that the leader was initially given, however, there was an estimated completion time for each operation broken down into six-minute intervals. In the past it would have taken a real effort to dig any information out of a mound of paperwork, but now key information would be more readily available from the computer.

According to Heide, "The leaders went berserk when they saw this." They suspected it amounted to a glorified time study. One leader expressed the widely held opinion that "it seemed to be a lot of information gathering by the company to keep a close watch on how long it took to build a die and on who was building it." A further fear was that the estimated times would quickly become the maximum allowable times. Donovan elaborates:

> We felt the time it took a diemaker to do a job would be put against the time it should have taken. If it took a guy four hours and 8/10ths, and it was supposed to have taken three hours and 7/10ths, he used an hour and 1/10th too much. You know, then they'll know exactly where the overrun or the underrun is.

The anxiety went beyond the timing of specific jobs. The additional worry was that all an individual's activities could be tracked, and even monitored over time. According to Donovan, "Through the use of the new time card they'd know the hours the guy worked, the job he worked on, the detail he worked on, and of course how long he took to do it." Another leader complained: "The information they gather this year, they're going to use next year when we build a similar die. They'll try to say this guy is fast and this guy is slow."

The diemakers felt that the unpredictable nature of their work meant that time study would be an extremely misleading way to judge performance. One leader argued:

> In our line of work, it can't be said to take this long period. If you build two identical jobs, it won't take the same amount of time and won't be done in the same way. Supposing you're putting a wear plate on, which is hard to do, and you break a tap? Where does that show on the time study?

Another leader stated: "We build everything in groups of one or maybe two dies. If a part comes through that has too much stock on it from the machining area, then the guy who has to grind it is going to look bad." A third diemaker objected: "A lot of factors are out of our control on the floor. Suppose we have to break a setup for a job the boss is trying to rush through. Where does that extra time get charged?"

Unbeknownst to the diemakers, however, the system did have some float built into it. If the total estimated hours to complete a series of details added up to 2,400, a somewhat larger figure, say, 2,700, was entered into the computer. These excess hours went into what was called a "honey pot" and were pulled out to take care of any unforeseen emergencies. A complex formula was developed to allocate expenses for the "honey pot" and this was also programmed into the computer. The planners, however, felt that if the workers on the floor saw these larger numbers, the actual construction time would magically expand to fill the allotted time. The result was the deletion of the float from the routing sheets that were distributed to the leaders.

Finally, many leaders complained that providing all this information to the system was a considerable burden for which they received no tangible benefit in carrying out their job. Some leaders complained of being stuck with two hours or even more of paperwork per day. The leader was responsible for recording how many hours each of the five or six diemakers working for him took for each operation as well as the job's status. But during rush periods diemakers were frequently shuttled back and forth between several jobs. As one leader put it:

> In order to stay on top of things in the way they wanted, we would have had to stand there with a clipboard and do nothing but watch people. One day you might be doing wear plates and then you put them aside and then you go back to them. We're all over the place. We started catching hell because a job would be signed

off in the system but someone was still working on it on the floor.

In response to all these objections, the systems group was willing to make some modification. After all, O'Brien was aware that "skilled workers are very zealous guardians of how they work and what they do." As a result, some worker input was sought following the initial presentations. O'Brien maintains:

> In the area where we needed something, we would sit down and talk with the machine operators about what they're doing. We would ask them what they needed back in terms of information. So they became pretty much aware of what was going on and, in fact, a lot of their ideas got put in. Something I learned was that the more involved they became, the better the system became and the better it was accepted.

This involvement, however, did not address a very central issue: Was the system merely a more effective way to gather information and improve operations or did it change the patterns of authority in the shop? And if the latter possibility were correct, who would win and who would lose? The leaders were being told that they would become more important with the new system. In fact, they were pivotal in entering information into the program and in some areas might be provided with more information than in the past. But the leaders' authority over scheduling, one of their most important tasks, would be sharply curtailed. In retrospect, O'Brien admits this would have been the case and gives an example of how this authority might have been undermined:

> There is no question that the leaders would be more closely scrutinized under the system. Say, for example, I'm a

> leader and I have four people working for me. Today, it just so happens that I have only enough work for three. It isn't in my best interests, as far as peer pressure goes, to raise the flag and have this extra person moved over to another group to become more productive. So I'll probably spread the work for three out among all four diemakers and be the nice guy. With the system, however, we would have immediately known that and told the foreman he had excess capacity in this area. The leaders naturally didn't like that.

Aware that their control was being eroded, the leaders were not about to cooperate in what they viewed as their own demise. They have a long history of effective collective action on the job, but in this case, they were somewhat hobbled by the fact the top leadership of the unit formally took a neutral position on the issue, maintaining the company had the right to run its business as it saw fit. A number of influential leaders, some of whom were officers in the unit, saw things differently and urged noncompliance with the system. They understood that the way in which they provided information could be an important source of leverage, particularly as the system was being implemented. As one leader put it, "You can't give them all that kind of information. That's your job. That's all you've got. You give it all to them and they're going to find a way to do it without you."

The first confrontation occurred over the time-reporting aspects of the system. This began as soon as the leaders saw the estimated times of the routing sheets they were given to familiarize themselves with the program, several weeks before it was due to become operational. Heide immediately knew that this could mean real trouble.

> Everyone was really negative about providing the company with information the union felt could be used against

> its members. It caused all kinds of havoc because they said, "What happens if this machine breaks down, you're only giving me a set amount of hours to complete this task and if I don't, then management's going to use this as a tool to write me up and/or fire me."

Heide then sought to have the monitoring aspects changed to avoid the sinking of the entire system.

> I said this is never going to work. The leaders are never going to fill it out accurately. They're going to fudge and cheat. You're never going to get the accurate history you want. And anyway it's not important to the system at this point in time to leave the hours in, so let's drop it. And after about two weeks of meetings and beating my brains against the door, O'Brien finally agreed with me and removed all the numbers from the printout.

While this was viewed as an important early victory for the diemakers, Heide points out that the estimated times remained in the computer. For the time being they wouldn't mean too much because no comparable times were being entered from the floor, but at some point in the future, when the rest of the system was debugged, the monitoring could easily be reinstituted. Heide maintains that the long-range plan was to computerize the time-keeping system in the plant and to enter this data in the same computer as TOPS.

Even with the operating times deleted, two objections remained: Control over scheduling was being transferred off the shop floor and the leaders saw no value to them from the system. As the plan was being implemented, the leaders were in a particularly powerful position. There were plenty of bugs and the leaders' cooperation was pivotal in removing them. According to Heide, "There were so many mistakes when the program first came out that it took two programmers full time just to correct errors so the system could keep running." One critical input was the daily transmittals, a form the leader filled out that told what his group had done and the status of the dies. The transmittal also recorded any deviation from the planned schedule so that future programs could be corrected based on actual experience.

There is some dispute about how forthcoming this cooperation was. Heide is an optimist. He maintains:

> Essentially you had some leaders who thought it was something new and worked with it real well and filled out the cards and you had other leaders that didn't like it and didn't do too much.

Overall, he insists that he had all the information necessary to run the system. Some of the leaders, however, maintain that the resistance was much stronger and more widespread than this. They contend that the leaders often would not fill out the paperwork or would not do it in an accurate way. Moreover, the leaders were supposed to meet with the coordinator every day they worked on the pilot dies. According to Donovan, this frequently didn't happen:

> The coordinator would come down to get this stuff and the leaders would say, "Hey, I ain't got time today. I've got to get my line-up out" or "I'm on my way out" or something.

In cases where there were real problems, Heide himself would fill out the transmittal or inspect the job. While he informed his immediate superiors about the dissatisfaction on the floor, he had the distinct feeling that it wasn't registering. Kuhn maintains the systems people felt that technical improvements and managerial firmness could overcome these or any other difficulties. He recalls that "these guys had boundless enthusiasm and boundless determination that they were going to break through what they felt was a curtain of resistance. They firmly believed that they were on the right track." The division-level managers, however, were making their decisions without a clear idea of the level of problems on the floor, according to Heide.

A second source of resistance came from the operating management in the plant. The management that had initially approved the proposal had been transferred. The new plant manager was close to retirement and not particularly interested in rocking the boat with new approaches to diemaking. In effect the plant was being run by the production superintendent.

The technical shortcomings of the initial plan were part of the problem. For example, TOPS was imperfectly linked to existing reporting and financial systems. Two reporting methods and a lot of double entries were therefore necessary, one for TOPS and another for the existing computer systems. More importantly, there was inadequate status reporting. According to Heide:

> There was not a good way of saying where you stood on a given die. The system didn't automatically flag overdue dies. In order to obtain this information, you had to go through and pick out the dates. If you looked at the whole program, with anywhere between eight to ten important dates for each of 650 dies, you're talking over 6,000 dates that would manually have to be scanned.

Many of these shortcomings would have no doubt been improved with time. But a far more serious problem was essentially political in nature: the system was perceived as revamping the existing distribution of managerial power. In the process of taking the roof off the tool and die building, the activities of the plant management were also exposed to their divisional superiors. For example, if the production superintendent reported that a die was 80 percent completed, but in reality some unforeseen problem meant that it was only 50 percent done, this discrepancy would be immediately obvious. The production superintendent would then lose the autonomy to correct the situation before the final due date. O'Brien contends:

> In a shop of this type, the person who is running it at the superintendent level is god. He can shuffle things around anyway he sees fit. If there is a problem on a job, he can throw more resources at it to bring it up to speed. The two superintendents saw the new system as having a tremendous impact on their authority. And in truth it was because they previously set the schedules and controlled the whole place.

Heide elaborates on this analysis:

> Initially the production superintendent tolerated the system because it was politically wise to do that. All the big shooters were interested in it, and several million dollars were committed. So he politically supported it thinking he could tailor the system so that he could have an even stronger hand over the plant than he had already. When he found out that this wasn't possible, that the system was going to be controlled from the people in the office building, then he started to get negative about the Rouge program.

Although unenthusiastic about the system, the ability of members of the "management team" to resist an approved program was more circumscribed than what could take place on the shop floor. Nonetheless, minimal cooperation could be deadly. O'Brien states:

> You can achieve cooperation in various degrees. They can say "I'll do anything in the world for you" as the spoken word, in front of everybody, but then there is the point where it comes down to actually doing something and working to make the system work. That is a different level of cooperation. Some of the plant managers stayed at the first level.

Minimal cooperation included quite a bit of internal criticism and sniping at the system. Heide describes some of it:

> The production superintendent discredited the system. Found every fault with it that could be found. He would find a date out of 60,000 dates that was one or two days off and boy oh boy. He just raised all kinds of hell, saying "I can't have this system. It doesn't give me factual information when I need it." And he just took every opportunity at a higher level to knock the program and say it was a failure and that the company was pouring millions of dollars down the drain.

This cool attitude at the top quickly worked its way through the entire management structure in the plant. The careers of managers in the building are dependent on their superiors in the operating hierarchy, not on the opinions of an outside systems group. If lower level supervisors detect hostility to a program, they know how to act. Moreover, some of the same concerns vis-à-vis job autonomy are present at all levels. O'Brien details the response of the first line supervisors:

> Again it was a situation where you had almost complete freedom within your area and now you are imposing a discipline on it. Although the foreman participated in developing the methodology and saying this is how it is supposed to be done, when you gave it back to them and

> told them this is how it's going to be done, they had a little resistance to that even though it was based on what they had told you in the first place. They had no idea it would be so finite and detailed. Some of them would then nitpick, saying "I have to do operation eleven before operation ten." And the way the system is built we said, "Go ahead, if you decide that's what you want to do, do it." But it was not perceived of as a guideline or a tool or whatever, it was perceived as a rigid set of instructions taking away their prerogatives.

As the months wore on, the disputes between the systems group and the operating managers became more and more intense. The leaders, however, were not directly aware of this widening gulf. Nevertheless, it had a substantial impact on them. Since the operating management at least implicitly opposed the system, they avoided cracking down on any lack of cooperation from the shop floor. Moreover, this conflict could be used to underscore the ineffectiveness of the program. Donovan recalls:

> If they had really been for it, we might have had more of a problem with our guys refusing. Because we've had other problems in there when the company has come down hard on leaders and diemakers and said "you're gonna do it." So we've had confrontations, but not on this.

Had the plant management enthusiastically supported the system, however, there might have been an even sharper response from the diemakers. Al Gardner, the current chairman of the unit, contends: "If they had pushed the system, resistance would have grown. It's a tradition that those leaders don't back down on anything."

Events finally came to a head about nine months after

the program was introduced on the shop floor. O'Brien describes the final denouement:

> What came to pass is that the plant manager began receiving conflicting status reports from the production superintendent and from the system. There was a die in construction that was about 200 to 300 hours from being completed for tryout and it was consistently behind schedule. We would consistently report this and we actually tracked the die on the floor. At the same time, the production manager was reporting to the plant manager in his weekly status meeting that this thing is on time or ahead of schedule. The question was does the plant manager believe the computer or his production team? After this was raised a number of times, we invited the plant manager to come out and look. Before that happened, we went out to check on the die. There were fifteen people working on it, standing around it shoulder to shoulder to get the work done. Now up to this point in time, we're saying it's two weeks behind and all of a sudden for three days you put fifteen heads on it doing whatever you want, trying to prove a point that no matter what the computer says we can do anything we want on the floor. At this point, I said, "I've had enough of this. I'm not Don Quixote. I don't keep tilting at windmills forever." So I recommended to management unless we get full cooperation we can't continue.

TOPS was withdrawn from the floor, but this was not the end of the systems group nor the last attempt to use computer technology to rationalize die production. In the next several months, toward the end of 1977 and beginning of 1978, several meetings were held with the divisional staff, the plant management, and the system group. At one of these meetings, a systems group staffer reportedly said, "We need to put a charge of dynamite into the tool and die building to let a little air blow through there so we can figure out exactly what's going on." At some point authorization was given to continue work on computer projects in this area, although on a scaled down basis. This was to lead to another attempt to introduce a modified TOPS system onto the shop floor in early 1979.

In the intervening year, a dozen or so computerized reporting systems were generated. These varied considerably in scope and complexity and some were never actually implemented. One system involved computerizing a hand-generated status report of die room operations. This report, called a "mini," was actually a massive document published every two weeks and updated daily. All plant management down to the level of foreman received a copy, and these sixty-five or so identical documents were used as a benchmark to help control production. Another program, developed but not implemented during this period, was the "golden rod" report. This was meant for divisional managers, informing them in general terms about the status of each line of dies under construction, particularly the percent complete and whether or not they were on target.

All this activity evolved into another attempt to bring computerization down to the shop floor in February 1979. A modified version of TOPS was introduced to three more limited areas of the plant: machining, Kellering, and Keller Buildup Unit (KBU). This last area is where the strongest resistance took place. It prepares the dies for Kellering, a complex machining operation that cuts the contours into the die on machines called kellers. In the KBU area, there were about four or five leaders on each of two shifts and a total of about twenty-four to thirty diemakers involved. KBU was the result of an earlier attempt to use work organization to gain increased control over diemaking. In the past, a leader would see a job through from start to finish. This area, however, concentrated only on the initial aspects of die construction. Many of the leaders thought this would lead to an erosion of skill. Donovan maintains:

> We viewed the area as a breaking up of our trade. It took a lot away from diemakers as a whole because it specialized the guy. You might be stuck for years just putting sections on.

The modifications to the original system were designed to correct some of the earlier shortcomings. For managers, in addition to the die routing, the program now spelled out the status of each die on a percent complete basis and flagged a number of important milestones in the construction process. For the leaders, the system now provided information they had requested the first time around. They had complained, for example, that they never knew where castings were and had to spend a lot of time physically tracking individual details through various operations. Now this data would be available.

Prior to the introduction of the revamped TOPS, the foremen in the KBU area would contribute to a daily report about the status of the castings for the "mini." They would, in part, obtain this information from a unique coding system the leaders themselves had devised, which made it possible to instantly learn the status of every job in KBU by looking at a single sheet. To the uninitiated it looked like some strange hieroglyphics, an amalgam of circles, triangles, arrows, and different colors. But to those familiar with it, each line and symbol told something about the progress of a casting. A line under the base of a triangle, for example, might indicate that one surface of a casting had already been machined.

The new version of TOPS was based on this method. Only now the leaders' "book" would be eliminated, and they would provide information directly to the computer system through a transmittal. This new system also provided more information about when new castings would arrive, whether a casting was to be machined at an outside supplier and, if so, when it was due back. The KBU leaders viewed this new approach as another thinly veiled attempt to undermine their position. While the foreman might be able to glean considerable information from this book, the new system demanded even more data. The leaders feared this data would be used to tell them what to do as well as to report on what they had done. Donovan recalls, "The leaders really got hostile. They viewed this as another way to edge them out or undermine what they did. But, it really backfired. I never thought the leaders would get that pissed off." The leaders were angered further because they felt their system had been stolen. According to Gardner:

> The leaders felt that they had developed this system and they didn't develop it for the company. They developed it for themselves and they weren't going to give it to the company. So that was one of their bitches. They didn't mind a foreman coming over and looking at the book, but they weren't going to give the whole book to him. Because it was their idea.

The same informal resistance of the first go-around with TOPS was now repeated. In addition, the "book" also disappeared, making it difficult for the supervisors to plug information into the system when the leaders refused. Instead, the leaders carried all the scheduling information on little cards in their shirt pockets. Donovan describes what took place:

> Naturally, the company wanted to know what happened to the book but when they inquired nobody knew anything about the book. What book? Now with the card in their pocket, if they wanted to show it to a foreman,

> they'd show it to him, but if they wanted to tell him instead, they'd tell him. But if they don't even want to tell him anything, it stays in their pocket. The foreman can't go in there and get it.

Heide maintains that although there was resistance in the KBU area, the machining and kellering leaders were more cooperative. He attributes this to the value of the information they were receiving from the system.

This time, however, the union also became involved. Al Gardner was president and Donovan was vice president. They formally challenged the system by making a demand that zeroed in on the issue of power. They demanded that the system be administered from the floor with the program coordinators all becoming hourly rated union members. Donovan recalls the demand:

> So, we discussed the system for quite a while and then we told the company, "O.K., we can buy this system if we do all the computer input and the coordinator's job becomes hourly." Since the leaders didn't actually enter any information into the computer, they didn't know where it was entered or how it was used. So we wanted to control what was happening upstairs. We felt we could do that by taking over those jobs. We felt the leaders could do it.

The company's rejection pointed to the real issue as being who would control the information and the scheduling, not whether or not computerization was going to take place. Heide indicates why the company rejected the proposal:

> The production manager didn't like the system, but he liked it a whole lot less with the union involved. From management's standpoint, if they had hourly people in

> those slots, then the union would have all the advance information about the program size, hours, and dollars that they didn't want the union to know.

In fact, Heide viewed the resistance of the leaders on the shop floor as not resistance against the system but as a lever to win this demand.

All of this culminated about two months after it began in a stormy meeting in the personnel manager's office with the production manager and the union present. Gardner describes his impressions of the meeting:

> We asked for a meeting with the production manager and we told him that we felt the system was being used to undermine the leaders and we wanted it stopped. He claimed that wasn't true and all the system was doing was keeping records. He denied it had the features we said it did. He claimed he had never seen anything like that. So at one point he calls Heide in the office and Heide being an honest guy describes the system as it is. After he leaves, the production manager slams down the book and says it will be removed. We kind of laughed about it afterwards because it seemed like it had been a show or something.

Coincidentally, Heide was transferred to another plant in the complex the following day. The move had been planned at least six months in advance but it gave a very different impression to the people on the floor. He describes his reasons for leaving the program:

> Politically, TOPS was a loser. There was no way to advance the system. I couldn't champion the cause and push it through the plant with the plant management telling me that it was the worst thing ever and the only

> reason they were allowing me to support it was because they politically had to. I felt that I was going to take the fall for its failure. I was a loser at the staff level because I represented the plant, and I was a loser at the plant because they needed somebody to hang it on and I was available.

Although the program was dead as far as the leaders were concerned, it limped on in a scaled down form with the foremen plugging in the required information. Finally, there was a small note in the minutes of a communications meeting in the tool and die building on July 13 and 14, 1981. The final epitaph to the system came under the heading "PLANT EFFICIENCIES HAVE RESULTED IN MANPOWER REDUCTIONS."

> Elimination of the IBM Tracking System (TOPS) and replacement with a more effective system eliminated one salaried employee.

Why did TOPS fail? The participants give different explanations. Heide maintains the first effort flopped because it wasn't an effective managerial tool. He goes on to say that the central issue was the poor working relationship between the company and the union before, during, and after TOPS. The technical failures of the first system, however, do not explain why the technically more sophisticated second version also failed. Moreover, the union officially cooperated the first time around. The resistance came from the leaders themselves, some of whom were cooperative with the company, but felt that their position was being undermined. O'Brien places the chief blame on lack of cooperation from the production management. While this was undoubtedly central in both attempts, it understates the impact of the resistance from the shop floor. A more complete explanation would include a combination of three intertwined factors:

technical flaws, discontent among production management, and shop floor resistance. The technical shortcomings in the system gave more leverage to the resistance of the leaders. In the second attempt to introduce the system, the union's official opposition also increased the power of the diemakers. Moreover, the production managers were able to point to this discontent as further reason why the system wasn't workable, thereby buttressing their opposition.

The system was designed with two very different purposes in mind: making operations more efficient by collecting and providing information and altering the relative power of divisional managers, production managers, and leaders in the factory. The system designers evidently gave little thought that the latter goal would spark a potent reaction on the shop floor. In any case, they felt that any dissent could be overcome. Instead, the system itself was overcome. By embedding in the technology an effort to change the relations of power in the workplace, the potential value of the computerization was lost.

Computers Off the Shop Floor: The Wider Context

So FAR I have been focusing on the uses of computer technology on the shop floor, exploring the ways in which jobs are automated and different parts of production are tied together and controlled. Off the shop floor, there are some equally profound changes taking place. In this chapter, I am going to examine the ways in which computers are used to revamp the design process, alter the structure of management, and expand production on a global scale. These seemingly disparate processes all share a common theme: Computer technology is the vehicle to unify the separate elements of the entire production system and extend managerial authority through all of them. As on the shop floor, the new shape of production is molded by the goals of those doing the transforming as well as by the technical possibilities available. In this case, however, the arena in which technology is revamped is much wider. Production is restructured from the designer to the assembly line, throughout the various layers of management, and geographically from Detroit to São Paulo. The changeover, however, is far from the smooth technical process its proponents would like it to be. Extending managerial control can cause sharp conflicts as those on the losing end fight back, and often effective operation is sacrificed to increased authority. Ironically, all of this once again is justified in the name of higher productivity and increased efficiency.

COMPUTER-AIDED DESIGN (CAD)

Few processes have been as totally revolutionized by computer technology as design. An explosive array of ever more sophisticated new systems, generally termed computer-aided design (CAD), electronically transform a designer's concepts into drawings on a TV-like screen, which can then be analyzed and transferred for use in a broad range of manufacturing applications. Consider the design of a simple bracket. In a typical CAD system, an engineer sits in front of a cathode-ray tube and sketches the bracket freehand using a special light pencil. The computer converts this rough sketch into the equivalent of a finished drawing, smoothing the lines, completing the bolt circles, and making countless other adjustments. When the basic design is complete, the engineer can blow up sections of it for a better view, shrink it for perspective, rotate it, compare it to designs of similar brackets, functionally test it, and even see how it fits in a larger assembly.

In early versions of the technology, engineers were limited to drawing in one color and in two dimensions as if they were working on an electronic drafting board. Now some of the more advanced systems enable the designer to manipulate three-dimensional solids such as cubes or spheres in much the same way that a child assembles a model out of blocks. The difference is that these simple shapes can be merged into new complex patterns on the screen. All of this takes place in color so stunning that the NBC peacock looks washed out in comparison. The functional purpose is to highlight the different features of a part or process.

Linking design to the computer opens up innumerable new possibilities. Rather than being liberated by these new powers, however, CAD in the service of maximizing profitability could cause the designer to become shackled by the technology. Before looking at the deployment of CAD in ways that demean design, let's first explore some of the potential benefits. Consider four of them: enlarging the number of design alternatives, enhancing the capabilities to make revisions, integrating design and manufacturing, and removing some limits of time and geography. The design possibilities are expanded because instead of experimenting with a half dozen brackets, an engineer now has the ability to let the computer run through several hundred alternatives. Calculations of stress analysis, thermal flux, volume, weight, and deformation can all be performed before the design ever leaves the screen. When all this is done, the relevant information describing the part is entered into a data base and this becomes the basis for all subsequent manufacturing operations. If an actual paper drawing is needed, an automatic drafting machine can produce one as fast as a high-speed plotter is capable of whizzing over the drawing surface.

There are also extraordinary advantages when it comes to making revisions of designs. In complex manufacturing industries such as aerospace, frequent design changes are routine and could create havoc if detailed drawings constantly had to be redone. Now the picture of the part can be called up on the screen, the necessary changes made, and a new drawing instantly produced. Moreover, the latest changes are available to any engineer plugging into the data base.

CAD exerts a powerful indirect influence on the organization of work: It facilitates removing the boundaries between different parts of production. On one level, CAD lays the basis for a vertical integration of operations from the designer's concept of a part to the point at which it is made. With conventional methods, an engineer would design the part, a draftsman would draw it, and a machinist would build it.

Now CAD is capable of translating a design directly into a part program that guides the cutting tool on an NC machine, eliminating all intervening steps between design and production. On another level, the central pool of information in the CAD data base serves to link together the various engineers on a job, a horizontal integration of the design function. On a large project, different teams of engineers feed their design work into the data base, which makes it instantly available to the other groups. Whether there are ten designers or a thousand, they all have the ability to view the same information at the same time. This makes it possible to work on different parts of an assembly in parallel without worrying about whether it will all "fit" when complete. Moreover, the design engineers are constantly able to interact with the manufacturing engineers, who determine how the part is made. If a particular product design is difficult to produce, the early input of manufacturing can trigger a search for other alternatives. Further, tooling engineers are able to look at the rough design of a part and begin engineering the jigs and fixtures that will hold it during production.

Telecommunications coupled with CAD removes the limits of time and space from design. As long as the designers are linked electronically, it hardly matters where they are located geographically. IBM, for example, utilizes its design and engineering talent from around the world to produce prototype computer chips. The corporation has a quick-turn-around-time (QTAT) production facility in East Fishkill, N.Y., that is linked by satellite to twenty-five design centers around the world where the circuit designs are actually developed on CAD systems. At the center, computers check the producibility of the designs and then provide instructions to make and test them. Typical three-month turnaround times are slashed to a week.[1] Even a complex manufactured product such as an automobile can be designed in this way.

In the design process, two central goals are cutting costs

and reducing lead times. According to a senior process engineer at the AC Spark Plug Division of General Motors:

> As exciting and novel and intriguing as computer graphics technology is, the real reason for its use, of course, is reducing costs and cutting lead time to improve our productivity, while at the same time improving the accuracy and quality of our designs . . . our savings ratios range from about 2:1 up to 20:1.[2]

Productivity gains of this magnitude are found at other companies as well. A project manager at Boeing, for example, maintains that "conservatively speaking, for selected applications, use of our computer graphics systems typically results in a 2:1 improvement over previous methods."[3] Improvements in productivity of between 2 and 30 have been reported by other firms, with some analysts accepting an overall figure of about 3.5.

This cost savings has led to a rapid diffusion of the technology into industry. Since the early 1970s, the sales of CAD manufacturers have mushroomed at a remarkable 40 percent annual rate. In 1981, the industry's revenues exceeded $1 billion. Most of these sales were of large sophisticated systems that can cost between $300,000 and $500,000. Simpler systems, that retail for under $100,000, are now coming on the scene, considerably broadening the technology's market potential. This price, for example, is within the possible range of some 65,000 U.S. firms that produce molds, patterns, castings, and mechanical designs and parts. Some analysts predict that the market for CAD systems could exceed $1.8 billion a year as early as 1984.[4]

In spite of the extraordinary potential of CAD to enrich design, the technology is being used in a way that is analogous to the restructuring of work on the shop floor, as a conduit for managerial power. Mike Cooley, formerly a senior design engineer at Lucas Aerospace in England and a leading trade unionist, has referred to the computer as "the Trojan Horse with which Taylorism is going to be introduced into intellectual work." CAD can become a vehicle to sever the designer even further from the realities of production; it may introduce new constraints into design, fragment work, allow skills to atrophy, and bring new isolation and stress to the job.

Since Taylor's day, the separation of planning from doing, "brain" work from "manual" work has characterized much of U.S. industry. The ideal has been for managers to determine what should be done, for engineers to calculate how it can be done, and for workers to do it. Along with this physical separation of function, a culture has enveloped engineering that increasingly leads engineers to perceive of themselves as an elite, separate from and superior to workers and the realities of the shop floor.

Conventional manufacturing, however, provides some limits to this separation. The nature of the production process itself compels engineers to have sustained contact with the shop floor. Design, for example, is an iterative activity, moving back and forth between the engineering office and the shop floor. An engineer must be in constant communication with the draftsman and the machinist, giving directions and listening to their feedback. After all, the drawing does not make its way to the machine automatically and even if it did, it usually requires instructions and interpretations once it gets there. The machinist may also have some useful new input about how to make the part, and then the design cycle goes through another loop.

With CAD, however, it is possible for the "drawing" to arrive at the machine without human intervention. Much of the engineer's traditional contact with the shop floor can begin to appear superfluous. All the data and more a designer could pick up at the machine appears to be available on the cathode-ray tube. The illusion arises that the model is, in fact, reality.

While some contact is obviously still necessary, the "feel" of production that only comes from a direct firsthand immersion in how parts are made can seem antithetical to the mystique of computerization. David Gossard of MIT, an authority on computer-aided design, expresses concern about this trend:

> The danger is the substitution of computation for experience. Unfortunately, computation is the holy grail that many are now stampeding toward. While this may make sense from a dollars and cents point of view, there is a danger of taking the numbers themselves all too seriously. There is a natural tendency to believe the numbers more than an experiment. As long as the engineering firm continues to do enough experiments, there may be no problem. But there are those who believe that all answers come out of a computer and they can get away from the reality of what engineering should be about.[5]

In the long run, this separation carries a high price. According to a report by a working party of the Council for Science and Society in Britain:

> Could human intelligence have arisen independently of the practical needs it served? The answer is undoubtedly no. Could modern science have developed in a society where craftsmanship and manual work were regarded as unbefitting the thinker? Again the answer must be no: a Greek philosopher could in principle have carried on experimental science through the agency of a slave; but those questions which could be answered only with a slave's assistance would have appeared unworthy of the philosopher's attention. So if, in industrial society, intellectual and manual work come to be finally and completely divorced, there must be a doubt whether this will not destroy the basis on which science and industrial development have themselves been able to flourish. There

> can, at the more personal level, be no doubt at all that to deny the experience of interaction between theory and practice is damaging to the development of the individual.[6]

An anecdote from a British aircraft factory illustrates the practical consequences of this trend. A young engineer designed the igniter for an aircraft turbine engine using the latest in CAD technology. The design was then produced on an NC machine tool. Somewhere in the process, a decimal point was accidentally moved one place to the right, resulting in a monster igniter, ten times normal size. When a worker lugged this giant igniter into the engineering office, the designer reportedly looked it over and commented on how good the design appeared, apparently unaware of the error. What is comical when it is this outrageous could be devastating in countless more subtle ways when the error is more difficult to detect and therefore might go into use unnoticed.[7]

While CAD opens up the possibility of countless new design alternatives, its use can also interject some sharp new limits into the engineering process. One limit is the designer's loss of an adequate frame of reference to judge the different alternatives the computer is posing. If the computer offers the choice between 600 brackets rather than 6, this is not a gain if the engineer has lost the ability to make a reasonable choice because of being so removed from the knowledge and experience of what a bracket actually does. As David Gossard puts it, "CAD won't change the fundamental judgment issues in engineering practice." Underlying the design of CAD systems are two very different possibilities. Howard Rosenbrock, one of Britain's leading engineers in the area of control technology, poses the choice as follows:

> The first is to accept the skill and knowledge of designers and to attempt to give them improved techniques and

> improved facilities for exercising their knowledge and skill. This demands a truly interactive use of computers in a way that allows the very different capabilities of the computer and the human mind to be used to the full. Such a route will lead to a continually evolving ability in the designer and a continued interaction between his ability and the techniques and equipment he uses.
>
> The alternative is to subdivide and codify the design process, incorporating the knowledge of existing designers, so that it is reduced to a sequence of simple choices. Thus "de-skilled," the job can be done by men with much less training and much less experience. In the areas affected, design skill will gradually die and there will be no effective dialogue between the operator of the system and the man in the Research Department who devised it and will modify it.[8]

In an increasing number of applications, the designer is required to choose from a "menu" of "optimized" alternatives. On simpler jobs, this approach can save quite a bit of time and repetitious work. It is widely used, for example, to design certain types of tooling in the automotive and aerospace industries. Jerome Green, an engineering group manager at the Fisher Body Division of General Motors, maintains that in his division there are "over 1,000 standards in 2D and 3D that permit us to design up to 90 percent of a tool like an erector set." But on more complex work, the extent to which creativity is throttled rather than capabilities expanded remains a real concern. Imagine, for example, Shakespeare trying to write King Lear with a wide array of "optimized" passages.

Subdividing and limiting the design process in this way interjects Taylorism into an activity that has historically been characterized by its open and creative nature. Mike Cooley projects that

> standard routings and optimization techniques may seriously limit the creativity of the designer, and the subjective value judgments would be dominated by the "objective division" of the system. That is, the quantitative elements of the design activity will be regarded as more important than the qualitative ones. There are also grounds for believing that overemphasizing the mathematical modeling of the design activity may result in abstracting the design activity from the real world.[9]

Gossard worries that

> computer-aided design tends to exacerbate the problem of specialization in engineering. It begins to have an insidious appeal to it that results in a tendency to restrict an engineer's activities. The requirements for integration that the system raises might mitigate this though firms that allow this to flourish do so at their own peril.

Although for senior designers, the application of CAD may open new horizons, for many others, intellectual speed-ups and mental hazards result in new forms of stress. Cooley writes:

> Where computerised systems like this are installed, the operators are subjected to work which is alienating, fragmented, and of an ever increasing tempo. As the human being tries to keep pace with the rate at which the computer can handle the quantitative data in order to be able to make the qualitative value judgments, the resulting stress is enormous. Some systems we have looked at increase the decision-making rate by 1,800 or 1,900 percent, and work done by Bernholz in Canada has shown that getting a designer to interact in this way will mean that the designer's creativity, or ability to deal with new

problems, is reduced by 30 percent in the first hour, by 80 percent in the second hour, and thereafter the designer is shattered! The crude introduction of computers into the design activity in keeping with the Western ethic "the faster the better" may well result in a plummeting of the quality of design.[10]

Frederick J. Norton, the manager of the interactive computer-aided design facility for the electronics engineering department at the Lawrence Livermore National Laboratory, concurs with Cooley:

> The first effect on the individual is the change in job content. An extreme example is where a drawing board is exchanged for a CRT screen. The act of design is inherently that of decision-making, but with interactive graphics the lower-level decisions are made by the machine. Two dangers exist. If previously one man performed both levels, the introduction of computer aids may remove the time when he previously let things mull over in the back of his mind. This may result in unacceptable high pressure on the individual. If, on the other hand, the two levels were previously separated between, say, senior and junior designers, the role of the junior designer may become an entirely clerical one, devoid of any decision-making content. This would result in considerable frustration if the task is still performed by a skilled person.[11]

COMPUTERS AND MANAGERIAL STRUCTURE

Increasingly, the division between computer-aided design and computer-aided manufacturing is becoming thinner. As we have seen, the information that mathematically describes a part in the CAD data base is electronically transformed into instructions that drive NC machines on the shop floor. This data flows together with other streams of information into what amounts to a "super" data base that describes all aspects of production—product design, engineering, production planning, inspection, and shop floor control. No factory begins to approach this level of integration today but the trend in this direction is very clear. Ultimately, timely access to this information potentially provides managers with a potent new control over the enterprise—unifying command in a way that reverses the fragmentation of the past.

Since Taylor's time, but before computerization, the drive for control over the enterprise has created a troubling paradox: Managerial control is fragmented by the very methods used to maintain authority on the shop floor. Splitting jobs into their smallest parts, for example, frequently means that management itself becomes atomized in the process of supervising these jobs. This is a hidden though very real cost of Taylorism to management. Dennis Wisnosky, the former chief of the Air Force's Integrated Computer-Aided Manufacturing program (ICAM), details the problem:

> In a valiant effort to get the job done manufacturing engineers continuously added complexity and controls. Then they divided the problem into small pieces, over and over again. As a result, many factories in many American industries are so complex that they appear to be unmanageable, labor forces seem to be out of control, and costs are all but unknown.[12]

Such organization lends itself to the establishment of fiefdoms on the level of middle management. While each of these subsections might strive mightily to meet its goals, the effort often comes at the expense of, rather than in concert with, the other parts of the organization. As Joseph Harrington puts it, "Optimizing the subdivision does not guarantee that the total organization will operate optimally."[13] Even the transmission of information becomes warped by these localized power centers. Harrington elaborates: "While information theoretically should flow freely from point of origin to point of use anywhere in an organization, all too frequently it is forced to follow the line structure of the organization up through the hierarchy to a common point and then down again to the destination."[14]

In contrast, computer technology provides a formidable tool for upper management to extend its control over production without splintering operations and thereby paying the price of its own disorganization. Jack Jessen, a tooling manager at Douglas Aircraft in Long Beach, California, claims that "the management cycle is compressed. For example, the engineer can design the model and the programmer can access it on a day-to-day basis. This gets people to work together when they may not have been used to doing so. It breaks down a lot of traditional barriers." Harrington goes well beyond this by stating:

> It now seems apparent that things are about to change—not incrementally, but radically. Fractionated management skills are being reintegrated and the new managers with their broader perspectives are directly controlling versatile machines capable of manufacturing diversified and customized products. The total manufacturing effort is being reintegrated into a responsive directable entity. It is a giant step and a step in a new directon.[15]

Harrington believes that the possibilities for managerial control over large-scale enterprises are comparable in scope and content to the domination of small-scale firms exercised by craftsmen generations ago. Prior to the industrial revolution, the producer and designer were the same person: a craftsman closely supervising production from the receipt of raw materials to the delivery of the finished product. For Harrington, an integrated team of managers will now perform this function. "Everyone above the foreman level will find himself working in an area of largely non-departmentalized decision making. The old barriers will dissolve under the irresistible demands of the new technologies, and communications will flow freely within the boundaries of the company."[16]

Whatever the technical possibilities, there are some clear social limits to this new integration of management. Chief among them are the middle managers, the leading candidates to become reintegrated into the organization. As seen with the introduction of the TOPS system, this group is particularly sensitive to the nuances surrounding changes of power. Realizing full well the potential of the computer, middle managers are in a unique position to defend their turf, especially when the new systems are being introduced. To them, information systems are a two-way street. While they may discover more about the operation, their superiors will also learn far more about how they are managing their department. Wickham Skinner, of the Harvard Business School, emphasizes the social dimension:

> Our efforts of decision making, communicating, scheduling, and supervising make up the infrastructure of our plants and these internal elements provide more resistance to change than the purely technological ingredients on which factory managers and engineers tend to focus.[17]

Frank Daley, formerly the manager of manufacturing development at General Motors, concurs:

> People have sensed that those in command of the computer were in power and that the changes do a lot more

> to the authority structure than appears on the surface. As a result, jurisdictional battles over who controls the computer have stood in the way of developing the technology. This arguing has slowed us down by at least five years.

This resistance, although it obviously has been very troublesome in the introduction of computers, does not mean that increased managerial control through computerization is impossible. As many upper-level managers are discovering, however, a sophisticated power-centered strategy is necessary to bring about the desired changes. As on the shop floor, computers and managerial desire alone are not sufficient to reorganize the situation. The response of those who are being reorganized is a central factor.

M. Lynne Markus, of MIT's Sloan School of Management, has completed an intriguing series of case studies that detail managerial problems in introducing new computer-based systems. A central theme running through these studies is the use of these systems to alter the power relations between different managerial groups, and the resistance this can induce from the losers. Markus comments on how serious this resistance can be:

> In spite of the increasing importance of organizational accounting and control systems and in spite of the increasing sophistication of the decision-making and information technology embedded in them, the record for successfully implementing these systems has been modest at best.[18]

Markus compares the goals of the information system with the existing distribution of power in the firm. Does the system enhance existing lines of authority or does it change power relations? The answer requires a careful look at the social and historical context at the time of introduction as well as an understanding of the technical features of the system. Markus delineates three areas in which the introduction of a computer system can challenge the current relations of power in an organization. The first concerns altering access to information, the second relates to the systems' use in modifying behavior, and the third deals with the symbolic aspects of information systems.

In a complex organization, access to and control over data can confer exceptional influence. "Information is power." One staff member had this to say about a colleague's role prior to the introduction of a new system: "Old Oscar has been the production controller for twenty years. He keeps all the numbers in his head, and he calls all the shots. No one can argue with him when he says 'we need this' or 'we only have that.' Oscar's vacations are events to be planned for months in advance."[19]

Oscar and other managers in a similar position are well aware that a significant share of their power grows out of this monopoly of information and are quite sensitive to the slightest changes that undermine their authority. Abstract pleas in the name of corporate betterment or social progress have little influence. In Oscar's case, he eventually accepted the new system when he developed alternative ways of preserving his organizational muscle. But, in those cases where the opposite occurs, managers are frequently reluctant to give in.

The second cause of resistance flows from the first: Access to information enables top managers more thoroughly to evaluate their subordinates and determine subsequent rewards and punishments. As Markus puts it "it thus becomes a matter of personal self-interest as well as organizational well-being for individuals to try to control the nature of the information collected and the choice of the measures designed into accounting and control systems."[20] If upper managers can receive detailed data describing day-to-day operations, then middle managers lose the autonomy to take corrective action before their superiors discover a crisis. A plant manager remarked:

> The problem is that we get evaluated against the forecast Sales makes for us. The fear is that we will be held accountable piece by piece, rather than for just the overall dollar figure. That we don't mind being held accountable for. But if they hold us accountable by the piece, and if Sales doesn't sell exactly the mix they predicted, we're in for it. The fear is that there is a lack of flexibility in the forecast.[21]

Finally, if the symbols of computerization promise one thing and the reality is different, then middle managers might drag their feet rather than cooperate. Some managers, for example, resent using word processors or even electronic mail terminals, feeling that this technology is associated with secretaries and an affront to their professional position. According to Markus:

> The organizational power structure is partially revealed through language and symbols, rituals and ceremonies; computer-based systems, likewise, have symbolic aspects and accompaniments. When the images of systems diverge from those of their organizational context, the existing structures are affronted.[22]

To cope with resistance, Markus advises managers that the way in which a computer-based system affects the distribution of power has central implications for how it should be implemented. In those cases where the new system reinforces the existing relations of power, user participation in the design and deployment can be a valuable tactic that lessens resistance to change. In those cases, however, in which the system is meant as a vehicle for changing the relations of power, user participation could be a serious error. Markus warns:

> If users are given a genuine opportunity to participate, they will try to change the proposed designs in ways which meet their needs to the exclusion of others. These attempts can lead to the failure of managers and systems analysts to achieve their "political . . . objectives."[23]

THE GLOBAL FACTORY

Computer technology combined with telecommunications makes it possible for management to recast the shape of production on a world scale. The power that results is unprecedented. There are extraordinary new capabilities to communicate, process information, coordinate operations, and direct production. Consider, for example, managers or engineers from a global firm sitting in front of electronically linked video display screens. They are able to manipulate the same data or interact with each other whether they are in the same room or separated by five or six thousand miles. A complex engineering change made on the screen in New York instantly appears on the screen in Tokyo, notice of a production foul-up in Mexico is immediately relayed to Detroit. Using the technology in this way removes key limits of time and geography from international management.

Paradoxically, this more-integrated managerial control leads to more-fragmented production. Multinational companies are already defining a new worldwide use of labor, based on where wage rates are lowest and local conditions most favorable, in what is rapidly becoming the global factory. While the technology does not revise the political realities around the world, it gives corporations the flexibility to exploit existing conditions to maximum advantage.

As globalization increases, there is a less direct relation between the financial fortunes of a manufacturing firm and the well-being of workers in any country in which it operates. In the United States, for example, a corporation's sales could be rising and its profits robust but the search for lower costs and even greater profits leads to production being transferred somewhere else. Whether the firm happens to be headquartered in the United States or not makes little difference. Computer technology enhances the ability to decentralize without the price of operations becoming disorganized.

This globalization of production lays the basis for larger and more powerful multinational firms. In some industries, markets are becoming increasingly international. To survive, a corporation must weather murderous competition and have access to large amounts of capital to support its worldwide activities. These turbulent conditions rapidly winnow out weak firms and lead strong ones to become more concentrated. The giants that remain will have unprecedented muscle in dealing with workers in any single country and could very well be capable of pitting entire governments against each other.

Nowhere are these developments more dramatic than in the auto industry. Detroit is in the midst of what Ford Motor Company board chairman Phillip Caldwell calls "the most massive and profound industrial revolution in peacetime history." In order to compete, the automakers are totally revamping their products and production processes. Moreover, they are pursuing a three-pronged strategy to slash labor costs: automation based on computers and microelectronics; givebacks and other concessions in wages and work rules; and outsourcing—the subcontracting of parts to lower-wage suppliers both in the U.S. and throughout the world.

The cost of the reconversion is staggering. The industry planned to spend over $80 billion in the five-year period beginning 1980, the cost of the Marshall plan in 1982 dollars. A good part of this investment will go toward reworking the factories of 5,000 suppliers and to revamping the automakers' 255 domestic plants. Detroit alone will be retooling or building from scratch 47 new engine and transmission lines and 89 assembly lines.[24]

Internationally, the changes may be even more far-reaching. Worldwide operations, of course, are nothing new for the automakers. Ford, the most advanced domestic automaker in this regard, has been building cars in other countries since the Model T. But these international operations were basically run as independent subsidiaries, owned by Detroit but controlled locally. The result was often fierce competition between different units and the needless duplication of design and manufacturing facilities. Today, however, Ford is well down the road to a worldwide car factory—a centrally directed, highly coordinated entity in which computer technology serves to harness resources for a single competitive thrust in the market. The change is symbolized by the "world car"—a standardized vehicle marketed throughout the world and built using global design and manufacturing facilities.

Before looking at the role of technology in the industry's restructuring, let's first examine the changing nature of the international market for automobiles. As is so often the case, the introduction of new technology is intertwined with other structural changes.

The size and character of the world market have shifted radically in the last several decades. The sectors of real growth are increasingly outside the United States and Canada, which accounted for 61 percent of noncommunist-market car sales in 1960. By 1980, this share had sunk to only 37 percent. During those two decades, North American demand increased 39 percent, while demand in Europe soared 188 percent and in Latin America exploded by about 600 percent. This trend is likely to continue. As Donald E. Petersen, president of the Ford Motor Company, put it: "While the North American market is still the single largest market in the world, the highly profitable 'growth' markets in the next decade and beyond will be those outside the United States."[25]

In the U.S. market itself, conditions have also shifted dramatically. Prior to the early 1970s, Detroit was effectively insulated from import competition by the nature of the product sold. Only the U.S. automakers were producing the ever longer, lower, and wider behemoths in the kind of volume necessary for high profitability. The periodic market penetrations of small imports such as the Volkswagen bug in the early sixties were viewed more or less as an annoyance—sometimes a major annoyance—but certainly not as a life and death challenge. All of this changed with the oil shocks of 1973–74. American consumers, confronted with soaring fuel prices, responded by demanding smaller cars in the showroom. The Japanese producers were already producing small vehicles and were in an excellent position to enlarge their market share, which they did. In the next several years, a volatile auto market and Detroit's lethargy slowed the changeover to smaller vehicles, which put the industry in a less than advantageous position when the Iranian oil crisis hit in 1979. Meanwhile, the Japanese automakers were building not only small cars but high quality and reliable vehicles. Detroit was more comfortable with its traditional values of style and performance. By 1981, the imports had won a 29 percent share of the market.

Slammed by the imports at home, Detroit is also facing increased pressure throughout the rest of the world. The Japanese, U.S., and European manufacturers are now slugging it out in virtually every market in the world and are doing so with similar types of vehicles. Half of Ford's business, for example, already comes from outside the United States. An important element of success or failure in these marketing battles will be the economies of scale possible with world production.

One important economy is the ability to harness global

engineering and managerial resources to design the vehicle. Here computers and telecommunications are key. Ford, for example, utilizes a $10 million computer center in Dearborn, Michigan, to coordinate worldwide product development. The center houses six general-purpose computer systems and a hundred special-purpose systems that are in use seven days a week. During the day, 5,000 Ford engineers and technicians across North America plug into these central computers through video display terminals and ordinary telephone connections. When it is night in Dearborn, Ford engineers in Spain, England, Switzerland, France, and Germany plug into the same data. The result is that designers on both sides of the Atlantic are able to work simultaneously on the same project.

The Escort, which Ford considers its first truly world car, was designed using this system. The basic design parameters of the car were determined at Ford world headquarters in Dearborn, after which teams of Ford engineers throughout the world developed the major components. German engineers, for example, designed the engine while Ford engineers in the United States designed the automatic transaxle. Constant access to the latest information from the other groups via the computer insured that all teams could work effectively in parallel. The transmission design, for example, could be modified to be compatible with any major change in engine specification. The *Financial Times* maintains that Ford saved $150 million in start-up costs and a considerable number of man years in engineering by designing one basic car for a number of markets in this way.[26]

This technology is capable of coordinating design work not only between different parts of the same company but also between an automaker and its subcontractors. The idea is to establish direct computer tieups between the firm and its design and tooling suppliers. Jerome Green, a Fisher Body group manager, stresses this point and indicates the scale of GM's activities in this area:

> "Communications in engineering" can be likened to an octopus with its many tentacles reaching out to form a highly effective network for the distribution of engineering information between Fisher Body central engineering, the plants, and vendors. Fisher Body has the largest combination of on-line and remote computer graphics systems in the world. We maintain enough computer power in-house for one and one-half moon shots.[27]

These computer hook-ups give corporate management new flexibility in where to locate engineering work, in transferring designs, and in controlling the supplier firms. Work at a supplier can be instantly monitored and if things aren't right, the design work removed and sent elsewhere. Green elaborates:

> We can, when necessary, move a 10 to 75 percent completed job from any of our local or plant design rooms to any other source for completion. This flexibility is a valuable aid in regulating and controlling our outside design load and vendor work force of over 1,000 engineers.[28]

Global design and a standardized product set the stage for worldwide production. The alternative strategy of exporting completed vehicles from one country, practiced with a great deal of success by the Japanese firms, may not be tenable politically in the long run. In the more advanced industrial markets, as imports become more successful, the pressure for some form of trade protection increases. In less-developed economies, the nascent auto industries are already protected. In particular, local-content laws mandate that a certain percentage of a vehicle's content—often exceeding 75 percent—be produced in the country in which it is sold.

General Motors and Ford are already pursuing an international strategy. In search of low wage rates, favorable government subsidies, docile or repressed unions, and to comply with local-content laws, Detroit has begun an ambitious campaign to secure new manufacturing sources throughout the world. With an increasingly standardized product, the U.S. producers are seeking the maximum economies of scale through the creation of export platforms—the concentration of production of a major component in one country and then the export of the part to assembly plants throughout the world. In this way, local-content requirements are also satisfied since the firm gains credits to import parts based on the value of what it exports. As Donald Petersen delicately puts it, "countries such as Brazil, Mexico, Taiwan and South Korea are offering skilled, low-cost labor and export incentives to firms willing to participate in their economic expansion."[29]

Mexico, for example, is becoming an export platform for engines, particularly for shipment to the North American market. The rush of producers to build or expand plants there has been nothing short of dazzling. Chrysler has completed a new engine plant in northern Mexico capable of producing over 300,000 4-cylinder engines a year; GM has built two new factories, one of which has the capacity to produce 500,000 6-cylinder engines and the other 100,000 diesels annually; Ford has constructed a 4-cylinder plant with an annual capacity of about 500,000 engines; Volkswagen has expanded an existing 4-cylinder facility to 300,000 units a year; and AMC along with Nissan is moving in a comparable direction.

This approach is often part of a dual-sourcing strategy in which each major component is manufactured in at least two different plants around the world. Dual sourcing seeks to reduce the risk of a single bottleneck or broken supply line disrupting worldwide production, and as will be seen, can be an effective weapon in a strike situation. In a declining auto market, dual sourcing can also leave workers in the higher wage countries extremely vulnerable. Chrysler, for example, manufactures the same 4-cylinder engine in Trenton, Michigan, as it does in the new Mexican plant. Saddled with excess capacity because of a weak market, the corporation chose to make its layoffs in Michigan because of lower costs in Mexico. The reason given was compliance with Mexico's local-content laws.

The automakers are also pressuring their larger suppliers to move with them to other countries both to satisfy local content and to reduce costs further. Thomas J. Busch, Bendix's vice president and group executive, Worldwide Business Strategy and Marketing, feels his firm is prepared:

> If the OEMs [original equipment manufacturers] come to us and say "Hey, we want the lowest cost product we can get, we don't care where in the world you get it for us, but get it for us," we can do that through the Bendix system. If they said, "We want to buy 30 percent of our master cylinders offshore and 70 percent at home, or 50–50, or whatever," we could say "Okay, we'll supply you 50 percent from the U.S. and 50 percent from our facility in Japan and have 100 percent capability in the U.S. in case there's a catastrophe on the oceans or something keeps you from getting your supply from Japan."[30]

Busch and others, however, complain that even this might not be enough and that the automakers' search for ever lower costs often disrupts the suppliers' operations. In some cases, different international plants of a supplier are pitted against each other for the lowest possible quote. Busch laments:

> The way they're doing it now is: Today they'll buy from Brazil because the exchange rate is right and the government gives them an incentive. Tomorrow the government changes its policy and they say, "Forget that, I'll go somewhere else." So they go to another facility, and

> all they're doing is running around and causing suppliers to respond to a very short-term contract, and we're saying, "Hey! Why don't we get together and do it on a long-term basis?"[31]

As a result of export platforms, both by the automakers and their suppliers, there could be a flood of imported parts for U.S.–assembled vehicles. In 1981, about 5 percent of the content of U.S.–produced vehicles was manufactured abroad, a percentage that could double or even triple by 1985. A March 1982 University of Michigan survey predicted that foreign content could soar as high as 26 percent by 1985 and then hit 36 percent by 1990.[32] These figures, which some see as high, nonetheless indicate the potential magnitude of the change. Even the more conservative Frost and Sullivan, a management consulting firm, looks toward 10 percent of the content of U.S.–built vehicles to come from Mexico and Brazil alone by 1985.[33] The Transportation Systems Center of the U.S. Department of Transportation has developed some projections based on public announcements by the automakers in 1980 and 1981. A rough estimate of the dollar value of these imports for 1982–84 comes to $2.5 billion, and this does not include the growing market for replacement parts, which on imported vehicles frequently come from the country of origin.[34] Including this category, the value of motor vehicle parts imported into the United States from countries other than Canada skyrocketed by 768 percent between 1970 and 1980.[35]

There are some important limits, however, to the globalization of production. For one thing, a certain "critical mass" of skills, local suppliers, and industrial infrastructure is necessary to build a major manufacturing plant. Some parts, such as transmissions, have tolerances that are so tight that automakers have been reluctant to build them in less-developed areas. Further, even with computers and telecommunications, long supply lines and distant plants can create problems of production control, larger inventories to compensate for the fragility of the supply lines, and higher shipping costs.

An indirect limit to internationalization is the health of the firm itself in foreign markets. Chrysler was forced to sell most of its more anemic foreign subsidiaries when it began its financial wobbling in the United States, thereby limiting its ability to source parts in its own plants elsewhere in the world. Even GM has had troubles gaining on the competition outside of the United States. A multibillion-dollar expansion program to bring out new models and build new plants abroad, announced in 1979, has fallen rather flat. GM is plagued by less-integrated operations than its rival Ford and by a top management that feels best when it is in the Midwest. Of the five members of GM's top executive committee, none speaks a foreign language and none has ever served the corporation outside of North America.

These difficulties notwithstanding, the trend toward worldwide manufacturing is plunging ahead. A key side benefit to an automaker is the ability to discipline labor by threatening to move production elsewhere in the world. Both Ford and GM used global sourcing as an effective lever to pry open their existing contracts with the UAW in late 1981 and early 1982. Peter Pestillo, vice president for labor relations at Ford, told the UAW that the decision as to the location of 17,000 Ford jobs would be made between November 1981 and September 1982, the normal expiration date of the union's contract.[36] The clear implication was that unless the contract was reopened and concessions given, many if not all of those jobs would be located outside of the United States. Roger Smith, the chairman of the board at GM, threatened "unless we can get a handle on excessive labor costs in our industry, there will be more plants shutting down—and more auto industry jobs going offshore."[37] Outsourcing continued to be a central contract issue in the 1984 negotiations.

The price tag for global production can be quite high, but those manufacturers too weak to afford it may fall by the wayside. The 110 independent car manufacturers in the world in the 1960s had already been reduced to 30 by 1980, and many analysts predict fewer than a dozen by 1990. The remaining firms are increasingly enmeshed in complex interlocking agreements. There are joint ownership arrangements, such as the purchase of a substantial amount of the stock of American Motors by Renault. There are joint production agreements for completed vehicles, such as the one taking place between British Leyland and Nissan in which a Nissan-designed vehicle will be built in British Leyland plants. And there are plans for the joint production of components. Volvo, Renault, and Peugeot, for example, are producing an engine jointly that all three firms will use. Donald E. Petersen summarizes the direction the industry is headed in:

> In the long run, I think that world trade in built-up vehicles will be largely replaced by trade in vehicle components. The car of the future will be a world car not only because of common design wherever it's sold, but also because it often will be built where it's sold, from parts that will come from many countries. Ultimately, the distinction between imports and domestic vehicles could very well become meaningless.[38]

Global production goes well beyond the automobile industry and is increasingly affecting all of manufacturing. There are not only world cars, but world machine tools, airplanes, construction equipment, television sets, and tractors. Ford's new Series 10 tractor evolved out of a five-year, $100 million international engineering program. North American engineers worked on designing 4-cylinder engines and the transmission while European engineers developed 3-cylinder engines. Assembly of the tractor will take place in Romeo, Michigan, and Basildon, England. Manufactured parts will flow in from throughout the world including Ford's transmission and axle plant in Antwerp, Belgium.

An important spin-off from the world car is the "world machine tool." Since about one third of machine tool production in the United States is destined for the auto industry and its suppliers, the restructuring of auto has enormous implications for the machine tool industry. The automakers are looking for standardized machines and systems that can be put to work wherever their production is located whether that means Russelheim, Germany, or Chihuahua, Mexico. For Ford's new Mexican engine plant, U.S. machine tool builders won much of the work but chose to source it to their foreign affiliates to take advantage of more favorable government financing or lower production costs or both. The Cross Company, for example, a Fraser, Michigan, machine tool builder, is building the plant's cylinder block machining equipment at its British subsidiary. As Glenn D. Babbitt, President of Motch, a machine tool builder, puts it: "By offering a broad line of machine tools in the world market, we can better aim our collective efforts toward a customer's world products."[39]

The global division of labor in the auto industry is obviously influencing the thinking of machine tool executives as well. Irwin H. Dennen, an executive vice president of the Ingersoll Milling Machine Company, defined the "world machine" for *Iron Age* in 1980 as a "machine that is designed from the ground up to be manufactured and sold in the world marketplace." Dennen sketched out three aspects of the world machine: standardization, design flexibility, and worldwide manufacturability. Standardization is a particular concern for machine tool builders because the United States is the only major industrial country that isn't using the metric system and it is vital that modules built in different areas be compatible. Design flexibility is critical so that the machine can be built with varying levels of technical sophistication depending on where it will be used. Dennen amplified:

> We also saw that the world machines of the future had to be designed to be installed, operated, and maintained in emerging nations, nations that simply do not possess the technology or sophistication to maintain a complex machine.
>
> At the same time, we saw that the machine had to be upgradable by options to the point where it would be at home in the most technologically sophisticated country.[40]

Finally, worldwide manufacturability is desirable for the machine tool builder so that subassemblies and components can be sourced in areas of lowest cost. William C. Gallmeyer, Chairman of the National Machine Tool Builders Association (NMTBA), has stressed that manufacturers should move some assembly and production operations out of the United States in order to take advantage of lower labor costs in other countries.[41]

Computers as Strikebreakers

THE ULTIMATE WEAPON workers can bring to bear against their employer is withdrawing their labor. In most industries, when a union strikes, production stops. The economic resources of the firm are then pitted against the staying power of the people on the picket line until the dispute is resolved. The union's leverage, however, is seriously eroded if the firm is able to continue operating while its employees are on the street. In a number of highly automated industries such as petroleum refining, chemical plants, and the telephone company, this practice has become normal operating procedure, obviously weakening the unions involved. In most industries, however, there are roadblocks to continuing operations, such as the need for hard-to-find skilled workers. The widespread introduction of computer-based machines and systems, however, removes some of these key limits. Computerization in many industries means that operations can be maintained with a less-skilled work force. It also becomes possible to transfer work outside of a strike location since telecommunications do not respect picket lines. Does this mean, then, that computer technology will be used to downgrade or perhaps eliminate labor's most effective weapon? In many cases, the possibility certainly exists. But, computers also provide labor with some potent new opportunities, such as the ability to paralyze highly integrated operations. Which scenario is enacted depends on the nature of the industry, the way the technology is designed and deployed, and the strategies workers and managers pursue in a given situation.

Consider a recent example of the use of computers in a strike. On August 3, 1981, members of the Professional Air Traffic Controllers Organization (PATCO) walked off the job, the culmination of a decade of bitter and often turbulent labor relations in the nation's air traffic control system. The Reagan Administration, determined to thwart a walkout of public employees, gave the strikers forty-eight hours to return to work or face permanent dismissal. The resources and muscle of the federal government were arrayed against a tiny union that threatened to cripple air transport throughout the United States. The stakes, however, went far beyond the air traffic system. A Wharton School monograph, *Operating During Strikes*, argues that "the federal government's handling of the PATCO strike is an investment in maintaining strike-free, sensible negotiations in federal service and in controlling labor costs," undoubtedly in both the federal and the private sector.[1] For PATCO, the battle ended almost before it was joined in a rout that included the firing of 12,000 air traffic controllers and the decertification of their union.

PATCO was weakened by a lack of public support, lukewarm aid or even hostility from other unions, its own inexperience, and a tough adversary. But what ultimately doomed the union was the government's skillful use of computer technology to keep air traffic moving, gutting the strikers' leverage. Soon after the walkout occurred, 75 percent of commercial flights were operating in spite of the fact that some 75 percent of the air traffic controllers were on the picket line. The centerpiece of the Federal Aviation Administration's strategy was "flow control," a computerized procedure to regulate departures and to space aircraft uniformly along air traffic routes, thus maximizing the use of airspace, facilities, and controllers.

The FAA's planning for a controllers' strike began in January 1980. For eighteen months thereafter, the agency worked to refine a plan for operating the nation's airways with minimum demands on controllers, including experiments with existing flow-control procedures. The earlier attempts sought to replace the controllers only during a strike; this evolved into permanently replacing the controllers once they went on strike. Some of the studies were shrouded in the greatest secrecy; computer tapes containing preliminary operating data were stored in locked safes at the leading FAA route-control centers across the country.[2] Even after a tentative agreement was reached with PATCO on June 22, 1981, the FAA continued to improve its contingency plans. Raymond Van Vuren, director of the air traffic service, maintains that "people think we have just been lucky in the way we handled the situation but the union has been telegraphing to us for the past two years what their objective was for 1981. We prepared for it."[3]

Computers reduce but do not eliminate the need for air traffic controllers, an occupation that remains labor intensive and skill based. According to a Rand Corporation report:

> Much, if not most, of a controller's time is spent on tasks that require distinctly human skills: negotiating flight-plan changes with pilots, vectoring aircraft around rapidly changing severe weather, deciding upon general operational configurations with other controllers, and the like. These tasks also require experience, maturity, and flexibility—the blips on those screens are, after all, real people who change their minds and make mistakes.[4]

As a result, supervisors were requalified as controllers during the planning period so that they could become the core of a group to replace any strikers. But one early study indicated that a work force largely limited to supervisors could maintain only about one third of normal operation, so clearly success depended on the number of controllers who remained on the job. As it happened, the 3,000 supervisors were supplemented by 5,700 nonstriking controllers and 1,000 military personnel when PATCO walked out, bringing the total to over half of the prestrike work force.

This turned out to be enough. With the new flow-control strategies in use and with nearly 10,000 controllers on the job, air traffic was disrupted within politically tolerable limits and the striking controllers were left essentially powerless. In pursuing this strategy, however, the government may have taken a major gamble on safety. In a complex system such as air traffic control where human life is at stake, built-in redundancy assures that if one aspect of the operation fails, the system itself will not. An important element of that redundancy is the skill and experience of the human operators. The administration's gamble has obviously paid off since a major air disaster has not occurred. But, by eliminating so many of the system's most experienced and talented controllers, the redundancy of the system may have been compromised. As one MIT expert on air traffic control, who prefers to remain anonymous, points out, "In this area, statistics are a rare event. You can have a ten-fold reduction in air safety and still have no accidents but that doesn't mean you want to do that."

An intriguing question is how much of the FAA's hard line at the bargaining table was based on confidence in its secret strike preparations. Or conversely, would PATCO have reacted differently had it known the full scale of the government's efforts to keep the system running? There is little question that the FAA maintained a hard line throughout the negotiations whether or not this was as a result of its contingency plan. Dennis Reardon, a negotiator for the union, describes his perception of J. Lynn Helms, the FAA chief, by saying, "Helms took four positions in response to our demands—No! No! No! and goddamn no!"[5] The feeling that the administration was being intransigent was also shared by John Dunlop, Secretary of Labor under Gerald Ford, as well as many other observers. According to Dunlop:

> What is absolutely without precedent, at least in modern times, is that [the Reagan Administration] has brought in no outside, dispassionate group to look at the problem. That ain't right. Also the Administration has decided . . . to leave no avenue of escape for the union. You just don't do that.[6]

Concerning PATCO's response, Reardon stated that if the government's preparations had been described prior to the strike, "We may have questioned the validity of the information, but we would have seriously considered it in making a final decision. . . . We're not interested in jumping off a cliff and onto a sword."[7]

With the strike over, the administration has turned its attention to rebuilding the air traffic system, criticized by many as being overburdened and outmoded even in the prewalkout days. On January 28, 1982, J. Lynn Helms announced a twenty-year program, costing between $15 and $20 billion, to replace the system's aging computers and significantly automate air traffic control.[8] An important question is to what extent the design of this program has been influenced by the government's strike experience and its turbulent record of labor relations. Whoever controls technological decisionmaking has the power to shape technology to conform with their desired goals.

Technically, there are a number of very different options available: These range from seeking total automation to giving controllers and pilots new tools that enhance human judgment. The FAA's proposal leans heavily in the former direction. Writing in the April 1982 issue of *Technology Review,* Hoomin D. Toong, a faculty member of the Sloan School of Management at MIT, and Amar Gupta, a postdoctoral research associate, describe an important aspect of the FAA's new program:

> Between 1989 and 1995, an automated en-route air traffic control (AERA) facility will be implemented to carry out normal routing and conflict-avoidance without controllers' intervention. . . . Such a system implies that the entire task of routing air traffic will be done with minimal human intervention, changing the controller's role from that of an active participant to that of a monitor. Only if the computer system shut down or judgments beyond the programmed instructions were required would direct human intervention be expected.[9]

The FAA denies that its automation program is an outgrowth of the strike, maintaining that it was begun in March 1981. This denial, however, does not indicate the extent to which the previous decade of labor strife was a factor in the minds of FAA planners. Helms does admit that the walkout heightened an awareness of the need to rebuild the system.[10] One FAA research and development executive told *Aviation Week and Space Technology* magazine that maintaining operations during the strike was in fact a test of the automation program.[11] The dispute certainly seems to have encouraged those forces desirous of eliminating controllers' jobs or at least minimizing their influence. In an editorial during the strike, *Aviation Week and Space Technology* commented:

> Yet one more positive result from the strike will be the acceleration of automation of traffic control. Automation

cannot affect the denouement of the strike, but the strike will unlock the decision and funding door to use avionic technology. . . . Few federal bureaucrats have the chance to fire 70% of their departments and replace the victims with junior, lower-salaried recruits—or with computers and black boxes.[12]

Toong and Gupta also feel that the strike will speed up the implementation of automation. "The government's expanded bargaining power means it can now implement such options to increase automation more readily and is moving aggressively to do so."[13]

The pressures to minimize the role of the controllers conflict with the development of the optimal technical alternatives. The Rand Corporation raises some probing questions about the role of the controller in any new system:

The critical question in designing the ATC [Air Traffic Control] system of the future is not really what *can* be done but what *should* be done. Exactly how much and what kind of automation should assist or replace the human controller? Should we strive for a system in which the machine has the primary responsibility of control and human expertise is used in a secondary, backup fashion? Or should men, in spite of their intrinsic limitations, retain primary control responsibility and utilize machine aids to extend their abilities?[14]

Rand then blasts the direction of FAA research and development for heading in the first direction:

The AERA scenario presents serious problems for each of the three major goals of ATC—safety, efficiency, and increased productivity. By depending on an autonomous, complex, fail-safe system to compensate for keeping the

> human controller out of the route decision-making loop, the AERA scenario jeopardizes the goal of safety. Ironically, the better AERA works, the more complacent its human managers may become, the less often they may question its actions, and the more likely their system is to fail without their knowledge. We have argued that not only is AERA's complex, costly, fail-safe system questionable from a technical perspective, it is also unnecessary in other, more moderate ATC system designs.[15]

Rand proposed an alternative called Shared Control in which the primary decision-making responsibilities remain with the controller but in which the operator has an increasing "suite of automated tools." The role of the controller would be expanded so that "he is routinely involved in the minute-to-minute operation of the system." The system itself would consist of a "series of independently operable, serially deployable aiding modules."[16] Whatever its technical merits, Shared Control would add to the responsibilities of an occupational category in which over two thirds of the existing members had just been dismissed.

The story of the confrontation between PATCO and the FAA underscores the potential importance of computers in labor-management relations in general and strike situations in particular. On the one hand, computer technology and telecommunications make possible central direction of far-reaching activities, concentrating enormous power into relatively few hands. The few dozen controllers in Gander, Newfoundland, for example, demonstrated their ability to halt virtually all trans-Atlantic flights for a number of tense hours near the beginning of the strike. Had a few more PATCO members joined the strike, the air traffic system of the entire country would have been tied in knots. On the other hand, complex computer systems often lend themselves to operation by a reduced and less skilled work force in an emergency situation.

The leverage of the air controllers evaporated because less-experienced workers could successfully take over the job.

A much different outcome resulted from a civil service dispute in Britain in the spring of 1981. In this case, a small number of workers used computers to bring an enormous bureaucracy to its knees. The civil service unions challenged the government in a pay dispute, not by pulling out hundreds of thousands of workers in a direct confrontation, but by withdrawing 3,500 workers who use computers to process Britain's national sales tax at tax collection centers. Great financial pain was inflicted with minimal resources in a few months by this devastating campaign of guerrilla warfare. In the first week of the strike, for example, 1,200 strikers reduced the government's revenues from a normal $550 million to $105 million, even though supervisors continued on the job.[17] Within a few months, labor stoppages had delayed between 25 and 45 percent of the government's sales-tax collections, forcing emergency borrowing: 370,000 payment checks piled up, creating a logistical nightmare.

A highly centralized computer system and a highly unionized work force with strong ties to related unions proved to be a formidable adversary for the government. According to Peter Jones, secretary of the Council of Civil Service Unions, "If you have half a million members and 1 percent on strike, the other 99 percent can happily go on paying their wages for them." In this dispute 525,000 civil service workers contributed about $2.10 a week in order to provide their striking colleagues with 85 percent of their pay. Management, however, has the power to redesign the technology, and in the aftermath of the strike, there have been widespread calls for decentralization of Britain's computer system. One spokesperson for a U.K. trade association, advising its members to use many small computers in place of a highly centralized system, coined the quip "an Apple a day keeps the union away."[18]

In more and more industries, the design of computer

technology provides management with some unique options. In operations as diverse as newspapers, insurance, aerospace, and automobiles, the ability of computers to continue production with a fill-in work force—often composed of people with fewer skills—can devastate a union effort. An early example of this was a bitter labor dispute at the *Washington Post* in the mid 1970s. Determined to free itself from restrictive work rules and to achieve lower manning requirements in the pressroom, *Post* management took a hard line at the bargaining table. At the same time, extensive preparations were made to keep the newly installed printing equipment running during a strike that was sure to ensue. These preparations included importing executives from papers with strike experience and sending fifty-five white-collar employees to a special school to learn how to operate the new presses, the automatic features of which were an issue in the dispute.

On October 1, 1975, the strike began. As 205 pressmen walked off the jobs, nine presses were smashed and one set afire. Whoever sabotaged the equipment undoubtedly suspected that this would be no ordinary walkout and that the very existence of the union was at stake. After suspending publication for a day, the *Post* used helicopters to ship printing plates to six nonunion plants in Virginia and Maryland, which would temporarily print the paper while its own presses were being repaired. After the machinery was fixed, a total of 35 managers and nonstriking workers performed the jobs of 205 press operators on the automated equipment. With the union's leverage sapped, the strike rapidly turned into a debacle for the pressmen. Not only was the strike lost, but only fifty-three pressmen were eventually hired back and all resigned from the union before resuming work.[19]

In the printing industry, what happened at the *Post* is hardly an isolated event. A. H. Raskin, the former labor writer at *The New York Times*, who has closely followed labor relations in the newspaper industry, maintains:

> New technology in almost every department has rendered obsolete the union's jealously guarded lines of craft monopoly; now a handful of executives and confidential secretaries with a modicum of special training can do everything necessary to produce a paper.[20]

Prior to the strike at the *Post*, the pressmen at the *Kansas City Star* went out. In May 1974, the dismissal of one union member escalated into the firing of ninety-eight union pressmen. Initially, there was broad support from other unionists on the paper and for the first week, 500 members of the other craft unions didn't cross the picket line. But the ability of managers and nonunion workers to put out the paper utilizing computerized typesetting and other automated techniques proved to be extremely demoralizing. The paper eventually was able to hire permanent replacements for the pressmen and slash manning requirements in half. In the same month, the pressmen at the *Morning News* and the *Times Herald* in Dallas also walked out. A central issue was management pressure to remove restrictions on the manning requirements for each press, long considered a fundamental union right. Once again, the existence of automation devices along with the fact other craft unions ignored a picket line combined to deal the pressmen a fatal blow. The number and severity of these defeats have created an atmosphere of fear and resignation among printing unions.

In the newspaper industry, however, managerial leverage is based on more than being able to print the newspaper during a strike. The paper has to be delivered as well. At the *Washington Post*, this proved to be no problem since the drivers were nonunion, but this is certainly an area of potential union leverage. The value of this broader support was illustrated in a 1978 strike of New York City's three principal dailies, the *Post*, the *Times*, and the *News*. In this convoluted and often bitter conflict, automation was a contributing factor to employer overconfidence as management went into negotiations. The industry's hardline position, particularly concerning control of the pressroom, resulted in a three-month strike, which the pressmen were able to survive. The support of other unions—especially the deliverymen's—proved to be more than a match for the publishers' ability to continue running the presses using automation.

Another confrontation between print workers and publishers, this one at the *London Times*, also illustrates both the power and the limits of high technology in a strike situation. Midway through a year-long lockout of over 3,000 workers in 1978–79, the paper's management sought to resume publication. The plan was to compose the paper in London, beam it to Frankfurt, West Germany, for printing in a nonunion shop, and then airlift it back to Britain. While the cost would have been prohibitive in the long run, the hope was that union morale would be sapped in the short run, thus forcing an end to the dispute.

The unions, however, had a different idea. They counterattacked by flying a delegation to Frankfurt; it successfully persuaded the city's Central Labor Council to take strong sympathy actions. Despite the fact the printing plant was nonunion, the ensuing uproar caused it to cancel its arrangements with the *London Times* after a single issue.[21]

Had PATCO been able to generate this kind of active support from other unions in the air transport industry, principally the pilots, the balance of power might have been changed considerably. The short-lived job actions that did take place overseas proved effective but difficult to sustain in the face of strong government pressures.

The advantages that computers give management in an office workers' strike were made clear during a 1980–81 dispute between Blue Shield of California and 1,100 members of the Office and Professional Employees Union (OPEU) Local 3 in San Francisco. As the 133-day strike began, the company adopted a carefully prepared contingency plan. This plan included assigning all available supervisors to computer banks in the claims-processing area, hiring and quickly training 350 new workers, and routing some claims processing to nonunion offices as far away as Los Angeles. The various offices were linked together through computers and telephone lines irrespective of picket lines. In addition, training the new workforce was made far easier because computers had been used to simplify tasks. As a result, Blue Shield asserted that it was able to maintain near-normal operations with far fewer workers. After the strike, the company refused to return 448 jobs to the main office.

Even manufacturing is affected. The power of skilled workers on the job is rooted in their skills, which in the past have been both difficult to replace and portable. As one machinist at Rolls-Royce Aircraft division put it, "If you have a dispute with the foreman, you take your knowledge home with you." This leverage is undermined, however, with numerical control. Much of the skill is embodied in the parts program, which is no longer under the machinist's direct control. During a walkout, experienced supervisors can instruct nonstriking employees, often with little machining background, on how to keep the machines running. Although the process can be inefficient and often produces considerable scrap, it can serve to pressure unions toward a settlement.

To utilize this capability fully, some companies have transferred experienced machine operators into supervisory and semi-supervisory positions before a strike. This is a particularly prevalent practice in the aerospace industry. At McDonnell Douglas in St. Louis, for example, nonunion managerial staffs are called Free Enterprise Personnel (FEP). During a strike in the late 1970s, Cas Williams, the president of the IAM local union, maintained that the company was able to operate at about 60 percent of capacity. Similar events have taken place at General Electric's giant jet engine plant in Evandale, Ohio. According to Homer Deaton, the president of the UAW local union in the plant:

> Almost every foreman in the NC area started out as a machinist and many of them are ex-union stewards. The company loads up supervision with machine operators because when we go on strike, they use foremen to set up the NC machines and then they bring in secretaries to run them.

Once the machining knowledge is embodied in the numerical control program, it becomes possible to transfer production from a struck plant to shops that are still working, regardless of whether they are across the street or half way around the world. This ability was demonstrated by General Motors when it brought out a new luxury model in 1975, the Cadillac Seville. It was designed in record time using the computer-aided design techniques available in the early 1970s. The computer also generated NC tapes to manufacture the car's body dies. When the Seville program was being planned, GM engineers decided to machine many of these dies at independent shops in the Detroit area rather than in internal GM facilities. Although the corporation may have preferred to go to nonunion shops, the independent tool and die plants with NC capacity for a job of this size are the largest shops and generally organized by the UAW.

A crisis occurred for General Motors when the UAW struck these plants in 1973. If the introduction of the Seville was not to be delayed, the die work would have to be done somewhere else. Tapes containing all the information necessary to machine much of the dies were sent to a General Motors plant in Flint that was not on strike. The enormous flexibility of NC allowed the Seville die work to be sandwiched between the already scheduled work and all projects were completed on time. GM admits to paying a $1 million premium for doing the work in this way but it was obviously worth the price.[22]

Without NC, it would have been impossible to transfer this project because of the large numbers of skilled workers necessary to complete it on time. Historically, the production of tools and dies has been a key bottleneck that the UAW has been able to use effectively in its struggles with the company. As it was, the leverage of the entire union was undermined in this case.

According to the Wharton School monograph *Operating During Strikes*, once a firm is successful in running a plant during a labor dispute, the practice is addictive.

> The fact that plant operation is a popular option for management with experience suggests that, once tried, plant operation tends to become an integral part of collective bargaining. This has been the case in the oil and telephone industries, is becoming the case in the chemical and newspaper industries, and may become the case in such new entrants to the field as the paper and hotel industries. Once used, plant operation may spread by virtue of example or force of competitive pressure, as in the oil and chemical industries. Once thus entrenched, plant operation seems unlikely to be dislodged until a new set of technological or institutional barriers to successful operation are created by either law or the ingenuity of the labor movement. No such barriers seem likely to emerge in the immediate future, but what lies beyond that limited horizon remains to be seen.[23]

The stakes for unions are obviously quite high. As the Wharton monograph puts it:

> The fundamental effect, if not purpose, of plant operation is to alter the balance of economic power in collective bargaining in management's favor by limiting the loss

> of revenue and profit resulting from a strike. In the short run, this enhanced bargaining power should enable management to secure a more favorable settlement of strike issues than would have been the case in the event of nonoperation. In the long run, this enhanced bargaining power should result in a series of more modest settlements, possibly to the point of seriously weakening the perceived effectiveness of a union or unions generally in representing employees.[24]

Where then does this leave unions? In simpler times, John L. Lewis—the fiery president of the United Mineworkers Union—told a president who sought to break a miners' strike that "you can't mine coal with bayonets." That still may be true, but computer technology allows more and more processes to be operated by supervisors and other fill-in workers during a labor dispute. Does this mean that John L. Lewis's strategy is outmoded and that the strike is finished as labor's ultimate weapon? No, but it means that unions will require broader strategies and more technical sophistication to use the strike as an effective weapon in the future.

While the labor movement's lack of ingenuity has been all too apparent in many instances, in some other cases innovative strategies have been the key to victory. One vital tool of workers, seeing their power eroded by computer technology, is to broaden the struggle to include those unions who retain considerable leverage. In the newspaper industry, for example, the New York pressmen were successful because the drivers refused to deliver the papers that obviously could have been printed. Moreover, international solidarity becomes a powerful weapon, as the workers at the *London Times* were to find out. In other cases, unions have revived older tactics with a new twist. In a recent dispute at the telephone company in British Columbia, for example, the Telephone Workers Union (TWU) knew that the company would be capable of continuing service if they went on strike. Instead, they occupied telephone company headquarters and offices throughout the province. The union, however, went beyond resurrecting the sit-in. Operations were continued "under new management"—the workers themselves—for five days. In a variant of this tactic, workers at the telephone company in Australia also continued working during a dispute in the late 1970s. But while they were providing service to the customer, they were refusing to process any bills for long-distance calls.

If new strategies are not developed by labor, its power could become increasingly eroded—first at the bargaining table, and ultimately in the society itself. The strike of the Communications Workers of America (CWA) against AT&T and the Bell System in the fall of 1983 is one more example of union leverage being sharply curtailed. While the strike disrupted certain services, such as telephone installations, the telephone company nonetheless was able to continue its core operations uninterrupted for over three weeks. Ironically, the strike occurred against a backdrop of near record profits for the telephone company and at a time when its relations with its union were among the most cordial of major U.S. industries. The company was obviously using power to protect its interests not simply in 1983 but for years to come in a deregulated market. If unions are weakened at the bargaining table in this way, then the erosion of the power of the labor movement in the society is not far behind. To the extent that power shifts in management's direction in negotiations, new industries, particularly high-tech industries, could become even more difficult to organize. None of this grim scenario for labor is inevitable. But in an age of high technology, "business as usual" is no longer a tenable strategy for unions.

A "Technology Bill of Rights"

THE EXTRAORDINARY CHANGES in production explored in this book are only the beginning of a radical shift in the way goods are produced and services are organized. The NC machine tools, flexible manufacturing systems, robots, and computer-aided design networks discussed in previous chapters are already working systems, cutting metal and transforming production throughout manufacturing. The next generation of these technologies will be far more sophisticated and even more widely used. The process of change, however, poses a fundamental dilemma for a democratic society: all of us will be affected by this revolution in production, yet few of us will have a hand in determining the basic goals the technology is designed to meet. Instead, an obsession with total managerial control is guiding the restructuring of production. Since increasing managerial authority requires decreasing worker input, computers and microelectronics are utilized to bring this about.

What happens if existing trends in the design and deployment of automation go unchecked? Current experience does not suggest a very promising future. In many systems, operators enjoy the most fulfilling and creative tasks when the machines don't fully work. Under normal circumstances, for example, an operator on a numerically controlled machine is only expected to load the part, press the start and stop buttons, unload the machine, and little else. In practice, far more of the worker's input is required, but this occurs in spite of the design. Considerable engineering talent and a great deal of expense are incurred to *avoid* using the extraordinary human capabilities available. As computer technology becomes increasingly sophisticated and reliable, more and more jobs could become routinized, electronically monitored, machine-paced, and, in general, more tightly controlled. Some jobs in batch production, because of their low volume and technical complexity, will not lend themselves to full automation. It is also true that other occupations requiring a high level of skill will be created as a result of computerization. The real issue, however, is not simply whether some old skills are no longer needed or some new skills are required but what the overall *direction* of technological development is and the reasons for that direction. While in some cases new skills may be necessary technically there is an unquestionable drive socially to minimize human input and establish tighter control in the workplace.

The factory may well prove to be a model for the way work is to be organized throughout the economy. The same conflict between managerial authority and worker autonomy is found in the office, the supermarket, the telephone company, and the warehouse. In these other areas, however, the process of change is accelerated. It took, after all, over one hundred years in the factory, from the time of F. W. Taylor until today, to rationalize, automate, and, finally, computerize production. Many offices, in contrast, will be transformed from a high degree of personal skill and control to super-automation virtually overnight.

Computers in the office, as in the factory, are increasingly tied into larger systems. These systems redefine the flow of information through the workplace and lead to the restructuring of all phases of work. The way individual tasks are revamped also remains important. Consider the word processor. Its many technical wonders, when used to correct errors or revise copy, can produce a better-looking document and save considerable effort. To a journalist or professional writer who controls its use, the word processor is invaluable. The benefits are also potentially great for secretaries or typists when they have some control over the technology, but when the word processor is used as a tool for rationalizing the office, the results can be a nightmare for those who have to use it. The Stanford Research Institute explains why: "Management has also found with word processing it can for the first time monitor the productivity of secretarial workers and thus gain some degree of control over this clerical function."[1] The theme is repeated by Wang, one of the largest producers of automated equipment for the office. An advertisement promises "A BUILT-IN REPORTING SYSTEM HELPS YOU MONITOR YOUR WORK FLOW." "You" in this case obviously isn't the secretary or typist involved. The reporting system "automatically gives the author's and the typist's name, the document number, the date and time of origin and last revision, required editing time, and the length of the document."[2]

One case study of office automation in an engineering consulting firm came to these conclusions:

> Interviews with typists indicated that the change from copy typing to word processing had reduced task variety, meaning and contribution, control over work scheduling and boundary tasks, feedback of results, involvement in preparation and auxiliary tasks, and communication with

> authors. . . . Management felt that closer control over typing work would increase productivity. The new work organization achieved this control, but appeared to restrict the typists' ability to exploit fully the potential benefits of the new technology.[3]

Other studies done in the United States raise similar concerns. The National Institute of Occupational Safety and Health (NIOSH), for example, studied the use of visual display terminals at five work sites in San Francisco in 1979: three newspapers, a newspaper agency, and an insurance company. NIOSH found a clear pattern throughout the research. "On every single scale of stress, physical or emotional or psychosocial, the clerical VDT operators scored the highest in terms of stress."[4] The stress did not primarily result from any inherent problems with the VDTs but from the way they were used. Managers and professionals using VDTs had a much lower stress level, resulting from their greater autonomy on the job, and a control group of clericals scored somewhere in between. When NIOSH examined the jobs of the clerical VDT workers they found "rigid work procedures, high production standards, constant pressure for performance, little control over their own tasks, and minimal participation or identification or satisfaction with their end product."[5]

It is ironic that computers and microelectronics should be used to create a more authoritarian workplace. They could just as easily be deployed to make jobs more creative and increase shop floor decision-making. Rather than pace workers, systems could be designed to provide them with more information about the production operation in general and their own jobs in particular. The technology could be used to bring the work under the more complete control of the people who do it rather than the other way around. These other possibilities, however, do not mean the technology is neutral. It is true that the basic components of new machines and systems—microelectronic hardware and computer software—can be deployed in a variety of ways. But once these building blocks are assembled into more complex designs, the bias of the designer is built in. The machines and systems themselves embody the relations of power in the workplace. As Langdon Winner puts it in his insightful essay, "Do Artifacts Have Politics?":

> The issues that divide or unite people in society are settled not only in the institutions and practices of politics proper, but also, and less obviously, in tangible arrangements of steel and concrete, wires and transistors, nuts and bolts.[6]

The use of technology to extend managerial power could have an adverse effect on productivity. Of late, slow growth in U.S. productivity has become a much-discussed issue. Most of the current debate, however, centers on which industrial sectors should receive available resources and on ways to increase the overall level of investment. Should managerial expertise and capital, for example, be channeled to new high-tech industries or to older smokestack firms? And, what measures are necessary to stimulate a more rapid diffusion of automated equipment? Automation is assumed to be not only the sure route to competitive success but unchangeable. Consequently, neither the nature of technological change nor the available alternatives are discussed. The assumption is that the more new machines and systems the better, regardless of their design, and policies are devised accordingly. Neglected in this approach is the *way* new machines and systems are developed and deployed, an especially important consideration since those approaches, which stifle human creativity, are incapable of utilizing the full potential of computers and microelectronics.

How, then, can technology be developed in a different

direction? A bold departure from current practice is needed. At the heart of this departure is a vision of technology that does not counterpose computers to people by relegating human abilities to the status of system static, an unfortunate variability that plagues efficient operation. Instead, automation can amplify rather than eliminate the unique qualities that only humans can bring to production, seeking to increase efficiency by tapping rather than destroying human creativity. In cases of conflict between higher productivity and more satisfying work, return on investment would be subordinate to human concerns as the chief guide to the development of automation. The realization of this vision involves a fundamental shift in the way decisions are made concerning the design, deployment, and use of new technology: It requires a broader public participation in determining the goals of technological innovation. Those affected by change, in particular, should have a central role in determining its direction.

No group has more to lose or gain from the process of technological change than the labor movement. On the shop floor, technology influences not only the satisfaction that workers gain from their jobs but the leverage they have in production. Off the shop floor, the central fear of many workers and unions is that not enough jobs will be available in the coming years. With rates of unemployment already high, an introduction of automation aimed only at maximum profitability could relegate many workers into a subclass of the permanently unemployed or underemployed.

In the face of the power and pervasiveness of computer automation, the strategies concerning technology formerly used by many unions are increasingly inadequate. Historically, management has claimed total corporate control over technological development. And, many unions have manifested a mix of cautious optimism and fatalism in their approach to technological change. These feelings are reflected in a two-year study done for the National Science Foundation, which concluded that "in the long run, it is willing acceptance followed by adjustment that constitutes the most common union reactions to technological innovation."[7]

The issues of most concern to unions have been job security and winning an equitable share of the economic benefits that new technologies can bring. Unions sought these aims by bargaining over the distribution of the product, not by challenging managerial control over technology. When it came to issues of technological change, unions largely reacted to managerial initiatives. In an uncertain economy particularly, job security and economic well-being remain at the core of any labor strategy. Technological change, however, can become the vehicle to undermine these traditional concerns. A clear example was the 1982 threat of Roger Smith, chairman of the board of GM, to introduce more robots in proportion to UAW wage hikes. Moreover, when it comes to conditions of work on the job, managerial desires are frequently already embodied in the design of the technology itself. To cope with this situation, unions are challenged to affect the course of technological development, not merely react to managerial initiatives.

An important first step in this direction has been taken by the International Association of Machinists (IAM). In the spring of 1981, the IAM convened an unprecedented conference of trade unionists, scientists, engineers, academics, and independent researchers to begin establishing a new framework to view technological change. A key result of this effort was a concept called "The Technology Bill of Rights." At its core is the principle that the introduction of new technology is not an automatic right of management but a process subject to bargained development. Before the IAM cooperates in the implementation of new machines and systems it wants to be sure that the technology will be used in a socially beneficial way. If certain preconditions are not met, the union asserts the right and responsibility to oppose the innovations. The document that came out of that conference follows:[8]

A TECHNOLOGY BILL OF RIGHTS

Proposed by the International Association of Machinists and Aerospace Workers

Preamble

Powerful new technologies are being poured into the workplace at a record rate. Based on the expanding capabilities and decreasing cost of computers and microelectronics, new forms of automation will leave few workplaces or occupations untouched. Robots on the assembly line, word processors in the office, numerical control in the machine shop, computer-aided design in the engineering department, and electronic scanners in supermarkets are only a few examples of the widespread changes that are taking place.

While such technologies offer real promise for a better society, they are being developed in a shortsighted and dangerous direction. Instead of benefits, working people are seeing jobs threatened, working conditions undermined, and the economic viability of communities challenged. In the face of these unprecedented dangers, labor must act forcefully and quickly to safeguard the rights of workers and develop technology in a way that benefits the entire society. Key to this is proclaiming and implementing a Technology Bill of Rights. This should be a program that is both a new vision of what technology can accomplish and a specific series of demands that are meant to guide the design, introduction, and use of new technology. This approach is based on the following assumptions:

1. A community has to produce in order to live. As a result, it is the obligation of an economy to organize people to work.
2. The well-being of people and their communities must be

given the highest priority in determining the way in which production is carried out.

3. Basing technological and production decisions on narrow economic grounds of profitability has made working people and communities the victims rather than the beneficiaries of change.

4. Given the widespread scope and rapid rate of introduction of new technologies, society requires a democratically determined institutional, rather than individual, response to changes taking place. Otherwise, the social cost of technological change will be borne by those least able to pay it: unemployed workers and shattered communities.

5. Those that work have a right to participate in the decisions that govern their work and shape their lives.

6. The new automation technologies and the sciences that underlie them are the product of a worldwide, centuries-long accumulation of knowledge. Accordingly, working people and their communities have a right to share in the decisions about, and the gains from, new technology.

The choice should not be new technology or no technology but the development of technology with social responsibility. Therefore, the precondition for technological change must be the compliance with a program that defines and insures the well-being of working people and the community. The following is the foundation of such a program, a Technology Bill of Rights:

1. *New technology must be used in a way that creates or maintains jobs.* A part of the productivity gains from new technology can translate into fewer working hours at the same pay or into fewer jobs. This is not a technical but a social decision. Given the pervasiveness of new forms of automation, the former approach is vital. The exact mechanisms for accomplishing this—a shorter work week, earlier retirement, longer vacations, or a combination—ought to be a prerogative of the workers involved. In addition, comprehensive training must be provided well before any change takes place to insure that workers have the maximum options to decide their future. Moreover, new industries that produce socially useful products must be created to insure the economic viability of regions that are particularly affected by technological change.

2. *New technology must be used to improve the conditions of work.* Rather than using automation to destroy skills, pace work, and monitor workers, it can be used to enhance skill and expand the responsibility workers have on the job. In addition, the hazardous and undesirable jobs should be a first priority, but at the discretion of the workers involved and not at the expense of employment. Production processes can be designed to fully utilize the skill, talent, creativity, initiative, and experience of people—instead of production designs aimed at controlling workers as if they were robots.

3. *New technology must be used to develop the industrial base and improve the environment.* At the same time corporate America has raised the flag of industrial revitalization, jobs are being exported from communities, regions, and even countries at a record rate. The narrow economic criteria of transnational companies are causing an erosion of the nation's manufacturing base and the collapse of many communities that are dependent on it. While other countries in the world have a pressing need and legitimate right to develop new industry, it is nonetheless vital that corporations not be allowed to play workers, unions, and countries against each other, seeking the lowest bidder for wages and working conditions. Instead, close cooperation among unions throughout the world and stringent controls over plant closings and capital movement are in order. In addition, the development of technology should not be at the expense of the destruction of the environment.

The implementation of a Technology Bill of Rights would obviously require profound changes at the collective bargaining table and in the political area. Unfortunately, the swift introduction of new technology won't wait until the proper mechanisms are available to deal with it.

The labor movement must seize the initiative. This means initiating and proclaiming a Technology Bill of Rights for workers and the society and making this a central vehicle to mobilize union members, organize the unorganized, and involve the community. In this way, corporate America can receive advance notice that the introduction of new technology is no longer the exclusive prerogative of management or an automatic process. Moreover, uses of technology that violate the rights of workers and the society will be opposed.

Instead of only responding to management actions, unions will seek full participation in the decisions that govern the design, deployment, and use of new technology. The goal will be machines that fit the needs of people rather than the other way around.

The "Technology Bill of Rights" points toward a different view of automation. Although it provides an alternative framework, it neither describes what a technology embodying human values would look like nor does it propose a specific strategy to transform the workplace. No advance blueprints are available and it would be impossible to reproduce the creativity and the excitement that the open and full participation of workers might generate. There are, however, examples of workers and unions that have challenged the current path of technological development. In the mid seventies, workers at Lucas Aerospace in Britain contributed an extraordinary vision of what worker participation in this area might look like. The process primarily affected the character of the products, but it would undoubtedly be similar for restructuring the production technology itself.[9]

Lucas Aerospace is a subsidiary of Lucas Industries, a British-based multinational company employing about 85,000 people worldwide. The Aerospace division makes aircraft components such as landing gear and electronic equipment. Between the late 1960s and the middle 1970s, the work force was reduced from about 18,000 workers to 13,000, a result of rationalization, new technology, and market shifts. At this point, the Lucas workers felt they had had enough. In response to these conditions, they developed an analysis of remarkable power and simplicity. They maintained that a deep contradiction existed between unmet human needs in the society and the economically enforced idleness of a highly skilled work force and some of the most modern plants and equipment in Britain. To resolve this contradiction, the Lucas workers demanded not only the right to a job but the right to produce products of value to the society. And they also sought a closer relation between *how* products are made and the purposes of production. They maintained it would be foolish to produce socially responsible products in an inhuman and alienating way.

After unsuccessfully soliciting the advice of experts, the union leaders and workers involved went to the 13,000 Lucas Aerospace workers located in seventeen plants. From this untapped human creativity, including 2,000 or so engineers and technicians, they developed a sophisticated and far-reaching plan to maintain jobs and create socially useful products. The initial plan encompassed five 200-page volumes, detailing the marketing prospects and production plans for 150 new products. Some of the plans involved expanded production of products that Lucas already produced, such as kidney dialysis machines. Other proposals detailed innovative new approaches such as a road-rail vehicle capable of running both on track and on unpaved roads, a feature of great value in many less-developed countries.

The plan was publicly presented in January 1976. Lucas management countered in April 1976 with a long, sophisticated response easily summarized in one word—no. Management even refused to negotiate at first, and although some bargaining later took place, the plan was never implemented. Nonetheless, it has had a significant and lasting value. The presence of an alternative model of production, generating widespread support within the British labor movement, prevented more layoffs from taking place. Each time the company threatened to close a plant, the Lucas workers would point to what they could produce in that facility. The plan served another purpose as well: It showed how the work force can develop technology in a way that meets their needs and the needs of the community. The roadblocks to action were primarily social, not technical.

Bargained development of technological change is already commonplace in a number of European countries. One of the earliest of these agreements was between the Norwegian Federation of Trade Unions and the Norwegian Employers' Federation. First signed in 1975, the agreement spells out important union rights in the implementation of new technology, and it sets up "data stewards," union representatives who monitor change. Further union rights were written into law in 1977. While such agreements underscored the need for broader public participation in technological decision-making, recent experience has indicated they are only as effective as their enforcement.

A further handicap to creating technological alternatives in the United States is a lack of union and community resources. Consequently, it is easy to assume that what exists is all that could possibly exist. The federal government, however, spends literally billions of dollars to insure that new machines and systems are developed in a certain way: the direction that industry wants them developed. Much of this money is channeled through the military. The Manufacturing Technology or Man Tech program of the Department of Defense, for example, will be spending about $850 million between 1982 and 1987. Since workers, unions, and the public are largely excluded from the planning process, this is a private appropriation of public monies. Ironically, for many people, these programs mean having their own tax dollars spent to automate their jobs or degrade their work environment. If 10 percent of these expenditures were devoted to the generation of human-centered technologies in local development facilities controlled by workers, unions, and public interest groups, an extraordinary diversity of systems would surely result. Ironically, many of these new approaches would undoubtedly be more productive than the current approaches that are pursued in the name of increased productivity.

The process of technological change does not take place in a vacuum. Without a more democratic control of the enterprise as a whole, the social control of technology will remain an illusion. Suppose, for example, that a union wins the right to decide how a robotic system is designed and deployed in a given plant. This "victory" gives the company a powerful incentive to move the plant. On the other hand, control over plant location pushes a firm in the direction of automation. Rather than deal with workers, the company only has to deal with machines, which do not demand their rights. Ultimately, the issues transcend collective bargaining and their political character becomes apparent. At stake is a more democratic structure of economic as well as political decision-making.

What is required to bring this about? One essential force is a revitalized labor movement willing and able to take on these larger challenges. While the risks are significant, labor is unlikely to hold onto its traditional gains unless they are taken. But the issues raised by this technological transformation affect more than labor: They are issues of central concern to the entire society. When work is electronically demeaned in the office or the factory, the repercussions carry far beyond the workplace. Work remains a central part of life; diminishing the human contribution on the job diminishes the quality of life off the job. And, a more authoritarian workplace could have a corrosive effect on democratic values throughout society. Whether we like it or not, the design of machines reflects social values as well as technical needs. The ideas and experience of those who are affected by new designs can help ensure that computerization will be a force that aids in liberating people rather than a vehicle for increased authority and control.

NOTES

1: INTRODUCTION

1 "The Speedup in Automation," *Business Week*, August 3, 1981, p. 60.

2 "Recession Even Hits Robots," *The New York Times*, January 12, 1983, p. D1.

3 "The Vision Behind Computervision," *Financial World*, January 1, 1982.

4 Barnaby J. Feder, "G.E. Offers Data Link to Improve Automation," *The New York Times*, March 31, 1982.

5 Edward E. Hood, Jr., "G.E.'s Commitment to Industrial Electronics" (Presentations to the Press and to Financial Analysts at the Hotel Pierre, New York, April 2, 1981), p. 2.

6 Philip Caldwell, "A Strategy for the Eighties" (Speech to U.S. Steel Good Fellowship Club, Pittsburgh, January 20, 1983), p. 3.

7 "Agreement Between General Motors Corporation and the U.A.W." (U.A.W. labor contract, September 14, 1979), p. 12.

8 *McAuto Services*, McDonnell-Douglas Automation Company, St. Louis, p. 41.

9 Edward E. Hood, Jr., "G.E.'s Commitment to Industrial Electronics," p. 2.

2: THE MACHINE SHOP

1 U.S. Department of Labor, Bureau of Labor Statistics, *Occupational Outlook Handbook, 1978–79*, Bulletin 1955 (Washington, D.C.: 1978), p. 38.

2 This section draws on material from: Hugh Aitken, *Taylorism at the Watertown Arsenal: Scientific Management in Action, 1908–1915* (Cambridge: Harvard University Press, 1960); Harry Braverman, *Labor and Monopoly Capital: The Degradation of Work in the Twentieth Century* (New York: Monthly Review Press, 1974);

Alfred D. Chandler, Jr., *The Visible Hand: The Managerial Revolution in American Business* (Cambridge: Harvard University Press, Belknap Press, 1977); Sudhir Kakar, *Frederick Taylor: A Study in Personality and Innovation* (Cambridge: MIT Press, 1970); David Montgomery, *Workers' Control in America: Studies in the History of Work, Technology, and Labor Struggles* (New York: Cambridge University Press, 1979); Milton J. Nadworny, *Scientific Management and the Unions, 1900–1932* (Cambridge: Harvard University Press, 1955); Daniel Nelson, *Managers and Workers: The Origins of the New Factory System in the United States, 1880–1920* (Madison: University of Wisconsin Press, 1975); Daniel T. Rogers, *The Work Ethic in Industrial America, 1850–1920* (Chicago: University of Chicago Press, 1974); Frederick W. Taylor, *Scientific Management* (New York: Harper Brothers, 1947). Taylor's three chief works are included in this single-volume edition: *Shop Management* (1903); *Principles of Scientific Management* (1911); and a public document, *Hearings Before Special Committee of the House of Representatives to Investigate the Taylor and Other Systems of Shop Management* (1912). Each work is paged separately.

3 Frederick W. Taylor, "The Principles of Scientific Management," in *Scientific Management*, p. 32.

4 Ibid., p. 104.

5 Ibid., p. 102.

6 Frederick W. Taylor, "On the Art of Cutting Metals," cited in Sudhir Kakar, *Frederick Taylor: A Study in Personality and Innovation*, p. 146.

7 Frederick W. Taylor, "Shop Management," in *Scientific Management*, p. 32.

8 Horace L. Arnold and Fay L. Faurote, *Ford Methods and Ford Shops*, cited in David Montgomery, "The Past and Future of Workers' Control," *Radical America*, November–December 1979, p. 10.

9 William J. Abernathy, *The Productivity Dilemma: Roadblock to Innovation in the Automobile Industry* (Baltimore: Johns Hopkins University Press, 1978), p. 24.

10 See Loren Baritz, *The Servants of Power: A History of the Use of Social Science in American Industry* (Middletown: Wesleyan University Press, 1960).

11 Montgomery, *Workers' Control in America*, p. 14.
12 This section is based on interviews done at the Rouge during 1980–82.
13 Edward A. Huntress, "Diesinking Today," *American Machinist*, May 1980, p. 152.
14 The Umpire, Ford Motor Co., and U.A.W., Request for Negotiation of "New Job" Classification Under Article VII, Section 23 (C), Case No. 39, 583-Appeal No. N-47761, PDH-267, October 10, 1979.
15 The Umpire, Ford Motor Co., and U.A.W., Request for Negotiation of "New Job" Classification Under Article VII, Section 23, Case No. 39, 583-Appeal Nos. N-4777, N-4780, N-4782, N-4787, PDH-266, October 10, 1979.

3: CRITERIA OF DESIGN

1 Machine Tool Task Force, *Machine Tool Systems Management and Utilization*, Technology of Machine Tools, vol. 2 (Livermore, Calif.: Lawrence Livermore Laboratory, University of California), p. 20.
2 Machine Tool Task Force, *Executive Summary*, Technology of Machine Tools, vol. 1, p. 17.
3 Raymond J. Larsen, Keith W. Bennett, and Barry H. LeCerf, "For Machine Tool Builders It's Been a Very Good Year," *Iron Age*, September 25, 1978, p. 33.
4 Donald N. Smith and Lary Evans, *Management Standards for Computer and Numerical Controls* (Ann Arbor: Industrial Development Division, Institute of Science and Technology, University of Michigan, 1977), p. 247.
5 Lamont J. Jenkins, et al., "Getting More Out of NC," *American Machinist*, October 1981, p. 185.
6 Ibid.
7 Joseph Harrington, Jr., *Computer Integrated Manufacturing* (New York: Industrial Press, 1973), p. 10.
8 "The Machine Tools That Are Building America," *Iron Age*, August 30, 1976, p. 158.
9 Russell A. Hedden, "NC Technology: Key to the Productivity Squeeze" (Numerical Control Society Keynote Address at the Sixteenth Annual Meeting and Technical Conference, Los Angeles, March 26, 1979).

10 Ibid.

11 Interview by author.

12 Ibid.

13 U.S. Department of Defense, Office of the Undersecretary of Defense for Research and Engineering, Defense Science Board, *Industrial Responsiveness* (Washington, D.C.: January 1981), p. 9.

14 George P. Sutton, "Training and Education," in *Machine Tool Systems Management and Utilization*, Technology of Machine Tools, vol. 2, p. 8.17-10.

15 Morris Moore, "Human Aspect of CNC" (Dearborn, Mich.: Society of Manufacturing Engineers, February 1979, mimeographed), p. 4.

16 Jack Thornton, "Automation Becomes Vital: A Shortage of Skilled Workers Is Forcing Firms to Turn to Computerized Machine Tools," *American Metal Market*, January 30, 1978, p. 18.

17 Alice M. Greene, "Lockheed-Georgia Advanced Technology," *Iron Age*, September 7, 1981, p. 115.

18 *American Machinist*, June 1982, p. 105.

19 "Replacing the Machinist's 'Little Black Book,' " *American Machinist*, May 1980, p. 6.

20 George H. Schaffer, "Implementing CIM," *American Machinist*, August 1981, p. 162.

21 Computer-Aided Manufacturing–International (CAM-I), Library Software Description, PS-76-PPP-03, Arlington, Tex., November 1977.

22 U.S. Air Force Integrated Computer-Aided Manufacturing, *ICAM Program Prospectus*, Air Force Materials Laboratory, Air Force Wright Aeronautical Laboratories, Air Force Systems Command, Wright-Patterson Air Force Base, September 1979, p. 8.

23 Ibid.

24 Ibid., p. 4.

25 "Cooperative Development and Enabling Technology Will Ultimately Produce CIM," *American Machinist*, June 1982, p. 173.

26 Quoted by Barry Rohan, "Rockefeller Sees Dangerous Split," *Detroit Free Press*, October 7, 1980, p. 36.

27 Stefan Aguren and Jan Edgren, *New Factories: Job Design Through Factory Planning in Sweden* (Stockholm: Swedish Employers' Confederation, 1980), p. 47.

28 Report of a Working Party, *New Technology: Society, Employment and Skill* (London: Council for Science and Society, 1981), p. 72.
29 Dr. Bernard Chern, "Production Research and Technology" (Paper, n.d.), p. 10.
30 Ibid.
31 Gideon Halevi, *The Role of Computers in Manufacturing* (New York: John Wiley and Sons, 1981), p. vii.
32 Machine Tool Task Force, *Executive Summary*, Technology of Machine Tools, vol. 1, p. 22.
33 Report of a Working Party, *New Technology: Society, Employment and Skill*, p. 76.
34 Ibid.
35 One of the few places where development work on "human-centered" systems is being carried out is at the University of Manchester under the direction of Professor Howard Rusenbruck.

4: NUMERICAL CONTROL: A CASE STUDY

1 Frank Lynn, Thomas Roseberry, and Victor Babich, "A History of Recent Technological Innovations," in National Commission on Technology, Automation, and Economic Progress, *The Employment Impact of Technological Change*, Technology and the American Economy (Appendix), vol. 2 (Washington, D.C.: Government Printing Office, 1966), p. 89.
2 Joseph Harrington, Jr., *Computer Integrated Manufacturing* (New York: Industrial Press, 1973), p. 64.
3 John H. Greening, "Build a 'Master Plan' for NC," *American Machinist*, March 22, 1971.
4 Cited in Walter S. Mossberg, "Are Pentagon's Planes, Ships, Tanks Getting Too Complex?", *Wall Street Journal*, March 13, 1981, p. 25.
5 Robert T. Lund, "Numerically Controlled Machine Tools and Group Technology: A Study of U.S. Experience" (Monograph, Center for Policy Alternatives, Massachusetts Institute of Technology, CPA-78-2, January 13, 1978), p. 37.
6 "The 12th American Machinist Inventory of Metalworking Equipment 1976–78," *American Machinist*, December 1978, p. 136.

7 Ibid., p. 133.

8 Anderson Ashburn, "The 1980 Machine-Tool Standings," *American Machinist,* February 1981, p. 93.

9 "The 12th American Machinist Inventory of Metalworking Equipment 1976–78," p. 137.

10 Paul Turk, "Computerization Moves Ahead at Caterpillar," *American Metal Market,* December 1, 1980, p. 9.

11 "The 12th American Machinist Inventory of Metalworking Equipment 1976–78," p. 135. "The 13th American Machinist Inventory of Metalworking Equipment 1983," *American Machinist,* November 1983, p. 113.

12 Anderson Ashburn, "The Changing Role of NC," *American Machinist,* November 1978, p. 5.

13 Herbert L. Wright, *Beginner's Course in Numerical Control,* Cincinnati Milling Machine Company, cited in Reinier Herman Kraakman, "Machina Ex Deo?" (Bachelor of Arts Thesis, Harvard University, April 1971), p. 19.

14 Ibid., p. 9.

15 Quoted in Jack Woodley, *Government Regulations Impact Prototype Tooling and Spur the Use of Numerical Control Machining,* Tooling Information Service, Detroit Tooling Association, May 1978.

16 Much of the material in this chapter relies on plant visits and interviews. Only interviews not conducted by the author will be cited.

17 What follows is based on personal interviews conducted at the plant and the following arbitration opinion and award: McDonnell-Douglas Corporation, St. Louis, and District Lodge No. 837, International Association of Machinists and Aerospace Workers, AFL-CIO, FMCS# 78K101702, Gr. No. 77–81, October 24, 1978.

18 David A. Price, "Putting Holes in the Tape: Management's Toughest NC Decision" (Society of Manufacturing Engineers Technical Paper, MS 78–158, Dearborn, Mich., 1978), p. 1.

19 Roger Tulin, "Taylorism on Tape: Numerically Controlled Machining in the Front Office and on the Floor" (Paper, December 13, 1978), p. 39.

20 Clinton S. Stanovsky, *Automation and Internal Labor Market*

Structure: A Study of the Caterpillar Tractor Company (Master's Thesis in Technology and Policy, Massachusetts Institute of Technology, June 1981), p. 69.

21 Leslie E. Nulty, "Case Studies of IAM Local Experiences with the Introduction of New Technologies," in Donald Kennedy, Charles Craypo, and Mary Lehman, *Labor and Technology: Union Response to Changing Environments* (University Park, Pa.: Pennsylvania State University, Department of Labor Studies, 1982), p. 119.

22 Ibid.

23 "1980 NC/CAM Guidebook," *Modern Machine Shop*, January 1980, p. 331.

24 Raymond J. Larsen, "Does Adaptive Control Still Promise Improved Productivity?" *Iron Age*, July 27, 1981, p. 57.

25 Richard A. Mathias, Wallace Boock, and Arthur Welch, "Adaptive Control: Monitoring and Control of Metalcutting Processes," in Machine Tool Task Force, *Machine Tool Controls*, Technology of Machine Tools, vol. 4, p. 7.13-2.

26 Quoted in Donald E. Hegland, "Numerical Control—Your Best Investment in Productivity," *Production Engineering*, March 1981, p. 46.

27 Edward J. Toton, "Numerically Controlled Die Build System" (Paper presented at Worldwide Manufacturing Productivity Conferences, General Motors Corporation, Warren, Mich., November 17–20, 1980), pp. 8–9.

28 Paul C. Miller, "Diemaking by the Numbers," *Tooling and Production*, October 1979.

29 Quoted in "Metalcutting: The Shape of Things to Come," *Iron Age*, December 17, 1979, p. 73.

30 Paul Prasow, "A Longitudinal Study of Automated and Nonautomated Job Patterns in the Southern California Aerospace Industry" (Report, Institute of Industrial Relations, University of California, Los Angeles, April 1969), p. 70.

31 Clint Stanovsky, *Automation and Internal Labor Market Structure*, p. 6.

32 Quoted in Paul Prasow, "A Longitudinal Study," p. 87.

33 Stanovsky, *Automation and Internal Labor Market Structure*, p. 76.

34 Ibid., p. 90.
35 Ibid., p. 93.
36 Ibid., p. 100.
37 Ibid., p. 90.
38 Ken Gettelman, "What Is the Right NC Programming Approach," *Modern Machine Shop*, July 1979, p. 97.
39 R. L. Hatschek, "NC Programming," *American Machinist*, February 1980, p. 121.
40 Albin F. Meske, "PROMPT NC Language Offers Simplicity and Speed," *NC/CAM Journal*, January–February 1979.
41 Ibid.
42 R. L. Hatschek, "NC Programming," p. 131.
43 Ibid., p. 133.
44 Quoted in Barry H. LeCerf, George A. Weimer, and Charles T. Post, "Metalworking's Future Manufacturing Systems," *Iron Age*, August 28, 1978, p. 108.
45 "(1) Paragraph 3, Recognition: Programming of Work for Tape-controlled Burgmaster Machine; (2) Classification of Operation of the Machine," Umpire Decision No. J-66, issued by GM Dept.—UAW, September 11, 1961, p. 4.
46 Ibid., p. 5.
47 Irving Bluestone, Letter to GM-UAW National Informational Subcouncils, June 19–20, 1978.
48 Information for this case was made available by Chris Laverty, a skilled tradesman and UAW committeeman at the Oldsmobile plant.
49 General Numeric advertisement in *American Machinist*, June 1982, p. 48.
50 Tree Machine Tool Co. ad in *American Metal Market*, February 13, 1978, p. 3.
51 Ray Milton, "Captive NC's New Role," *Modern Machine Shop*, February 1980, p. 107.
52 This case is based on Leslie E. Nulty, "Case Studies of IAM Local Experiences with the Introduction of New Technologies," pp. 126–30, and additional materials and interviews provided by the IAM.
53 David C. Gossard, "Analogic Part Programming—Automation for the Small Job Shop" (Paper, Massachusetts Institute of Technology, n.d.), p. 5.

54 Lawrence O. Ward and Arthur Welch, "Part Programming," in *Machine Tool Controls,* Technology of Machine Tools, vol. 4, p. 7.7-2.

55 Chris A. Laverty, "Numerical Control and the Machinist" (Paper, March 7, 1979), p. 13.

56 Ibid.

57 Ibid.

58 Ibid.

59 Ibid., p. 14.

60 J. G. Swim, "Centralized NC Pays Off," *American Machinist,* 1970, p. 95.

61 Cited in Albert E. DeBarr, "Safety, Noise, and Ergonomics of Machine Tools," in *Machine Tool Mechanics,* Technology of Machine Tools, vol. 3, p. 8.17-7.

62 Wright, *Beginner's Course,* p. 10.

63 Lund, "Numerically Controlled Machine Tools," p. 31.

5: THE COMPUTERIZED FACTORY: ON THE SHOP FLOOR

1 Quoted in *Iron Age,* December 17, 1979, p. 66.

2 M. M. Barash et al., "Optimal Planning of Computerized Manufacturing Systems (CMS)," NSF Grantees' Conference on Production Research and Technology, West Lafayette, Ind., September 27–29, 1978, p. E-1.

3 White-Sundstrand Machine Tool Company ad in *American Machinist,* November 1983, p. 233.

4 Steven Ashley, "Lockheed Plans Advanced DNC," *American Metal Market,* June 14, 1982, p. 9.

5 Joseph Harrington, Jr., *Computer Integrated Manufacturing, Industrial Press* (New York, 1973), p. 88.

6 The description of the Messerschmitt-Bolkow-Blohm factory is based on Roon Lewald, "Flexible Systems Makes Aircraft Parts," *American Machinist,* March 1981, p. 107. There are some important differences between FMS systems. For one thing, the product ranges from small items such as light-alloy transmission casings for electric generators to huge parts such as heavy cast-iron transmission cases for tractors. On some systems, the loading and unloading of parts at the machine is fully automatic; in other systems, such as Messerschmitt, this is still done by hand. The most advanced systems have random processing, a capability that allows parts to be processed in almost any order, which means that if a machine is down, parts can be routed to other machines. In some instances, the FMS comes in all at once as a unit while in other cases it is built up over years. At Messerschmitt, for example, the first machines of the $50 million system were purchased in 1971, most of the machines were brought in during 1976–77, and the central computers came last of all.

7 In this chapter, only those interviews not done by the author will be cited.

8 Donald E. Hegland, "Flexible Manufacturing—Your Balance Between Productivity and Adaptability," *Production Engineering*, May 1981, p. 41.

9 The description of these four cases is drawn from: Donald Gerwin and Jean Claude Tavondeau, "Uncertainty and the Innovation Process for Computer Integrated Manufacturing Systems: Four Case Studies" (Paper, March 1981).

10 Quoted in Donald E. Hegland, "Flexible Manufacturing—Your Balance Between Productivity and Adaptability," p. 42.

11 Raymond J. Larsen, "Taking the Labor Out of Manufacturing at Cincinnati Milacron," *Iron Age*, September 28, 1981, p. 104.

12 Paul S. Borzcik, "Flexible Manufacturing Systems," in Machine Tool Task Force, *Machine Tool Systems Management and Utilization*, Technology of Machine Tools, vol. 2, p. 70.

13 Moshe M. Barash, "Integrated Flexible Manufacturing Systems" (Paper, September 1976), p. 2.

14 Bruce Vernyi, "Building Tools the Japanese Way—In the US," *American Metal Market*, December 14, 1981.

15 Lewald, "Flexible System," p. 107.

16 Melvin Blumberg and Donald Gerwin, "Coping with Advanced Manufacturing Technology" (Paper, School of Business Administration, University of Wisconsin, July 1981), p. 2.

17 Borzcik, "Flexible Manufacturing Systems," pp. 70–71.

18 R. A. Berdine, "Caterpillar's DNC System 2½ Years Later" (Paper presented at Fifteenth Numerical Control Society Annual Meeting and Technical Conference, McCormick Inn, Chicago, April 9–12, 1978), p. 115.

19 Melvin Blumberg and Donald Gerwin, "Coping with Advanced Manufacturing Technology" (Paper presented to conference on Quality of Work Life and the '80's, Toronto, August 30–September 3, 1981), p. 6.

20 Ibid., p. 12.

21 Jack Hollingum, "An Automated Machining System Three Years On," *The Engineer*, May 12, 1977, p. 37.

22 Jack Thornton, "DNC System Boosts Machine 'Uptime' for Caterpillar," *American Metal Market*, May 1, 1978, p. 10.

23 Jack Thornton, "Allis Chalmers Retools on FMS," *American Metal Market*, May 10, 1981, p. 15.

24 M. M. Barash, "Hardware of Computerized Manufacturing Systems" (Paper, Purdue University, December 1976), p. 8.

25 Blumberg and Gerwin, "Coping with Advanced Manufacturing," p. 7.

26 See Melvin Blumberg and Donald Gerwin, "A Survey of Perceived Job Characteristics, Work Attitudes, and Working Conditions on Allis Chalmers' DNC Line" (Paper, School of Business Administration, University of Wisconsin, December 1980).

27 Blumberg and Gerwin, "Coping with Advanced Manufacturing," p. 12.

28 "Flexible Manufacturing: The Technology Comes of Age," *Iron Age*, September 7, 1981, p. 82.

29 Joseph Jablonowski, "Just How Many Robots Are Out There?" *American Machinist*, December 1981, p. 120.

30 Henry Scott Stokes, "Japan Looks to Big Gains in Robot Use," *The New York Times*, December 26, 1980.

31 "Estimated Programmable Robot Usage by Automobile Manufacturers and Suppliers in 1980" (Table, Transportation Systems Center, U.S. Department of Transportation, April 29, 1981).

32 Thomas O. Mathues, "Robots Working for Increased Productivity at General Motors" (Talk presented to Robotics International of SME, Chicago, September 30, 1981), p. 5.

33 T. W. Netherton, "Westinghouse Involvement in Industry Automation" (Speech to Robotics VI, Detroit, March 2, 1982), p. 7.

34 Robert B. Aronson, "Let the Robot Do It," *Machine Design*, November 27, 1975, p. 54.

35 Quoted in Steven Ashley, "Robots that 'Feel' Are Needed to Automate Many Plant Tasks," *American Metal Market*, June 29, 1981, p. 10.

36 Jack Thornton, "GM Pinpoints Its Robotic Needs," *American Metal Market*, March 22, 1982, p. 10.

37 Interview in *Production*, March 1981, p. 119.

38 Joseph Engelberger, Society of Manufacturing Engineers Technical Paper, MS 74–167, 1974.

39 Frank A. DiPietro, "Robotics, A Divisional Perspective—Past, Present and Future Applications" (Paper presented at Worldwide Manufacturing Productivity Conference, General Motors Corporation, Warren, Mich., November 17–20, 1980, p. BB-24.

40 Andrew Pollack, "GM Joins in Robot Venture," *The New York Times*, March 26, 1982.

41 Quoted in "Robotics Has Turned the Corner, Says Polcyn," *American Machinist*, December 1980, p. 81.

42 Jack Thornton, "Robots—Is Long-Expected Boom Under Way?" *American Metal Market*, October 12, 1981, p. 12.

43 John Holusha, "GM Shift: Outside Suppliers," *The New York Times*, October 14, 1981, p. D1.

44 Joseph F. Engelberger, "Designing Robots for Industrial Environments" (SME Technical Paper, MR76–600, 1976), p. 6.

45 Joseph F. Engelberger, "Robots and Automobiles, Applications, Economics, and the Future" (SAE Technical Paper, 800377, 1980), p. 6.

46 Kenichi Ohmae, "Steel Collar Workers: The Lessons from Japan," *Wall Street Journal*, February 16, 1982.

47 "Automated Machine Loading: Some Whys and Some Hows," *Production's Manufacturing Planbook*, 1978, p. 139.

48 Unimation Inc. ad, *American Machinist*, March 1978, p. 108.

49 Auto-Place ad, *Robotics Today*, Fall 1979, p. 40.

50 R. C. Beecher and Robert Dewar, "Robot Trends at General Motors," *American Machinist*, August 1979, p. 71.

51 Ibid., p. 72.

52 SOFI (Soziologisches Forschung-institut Göttingen), "Social and Organizational Aspects of the Introduction of Robots into Industrial Production," 1979, p. 8.

53 Ibid., p. 9.

54 See J. R. Houser, "CAM: Quality and Productivity" (Presentation to Nineteenth Annual Meeting and Technical Conference, Numerical Control Society, April 1982, Dearborn, Mich.); and Joseph Jablonowski, "Building Transaxles for Ford's 'World Car,'" *American Machinist*, June 1980.

55 See Thomas G. Gunn, "The Mechanization of Design and Manufacturing," *Scientific American*, September 1982, pp. 121–23.

56 Bryan H. Berry, "Detroit's Auto Industry Wrestles with Machine Downtime," *Iron Age*, September 16, 1981, p. 31.

57 Charles F. Carter, Jr., "Machine-Tool Technology in the '80's," *American Machinist*, December 1979, p. 82.

58 Gary J. McCoy and S. M. MacMillan, "Measurement and Control of Material Handling Equipment" (Paper, n.d.), p. 1.

59 Ibid., p. 13.

60 See William T. Lesner and Kevin G. Hughes, "Plant Floor Information Systems via Programmable Controller/Computer Networks" (Paper, n.d.); and William T. Lesner, "Computerized Machine Monitoring and Maintenance Dispatching System" (Paper presented at Worldwide Manufacturing Productivity Conference, General Motors Corporation, Warren, Mich.).

61 Hulas H. King, "DNC Management Data Reporting System" (Presentation to Numerical Control Society, Sixteenth Annual Meeting and Technical Conference, Los Angeles, March 25–28, 1979), p. 254.

62 Thomas J. Drozda, "New Controls on Uptime Boost Man/Machine Output," *Production*, September 1979, p. 78.

63 Lesner, "Computerized Machine Monitoring," pp. 1–14.

64 Ibid., pp. 1–16.

65 George G. Fenton, "Shop Reporting and Incentive Pay," *American Machinist*, April 1979, p. 142.

66 "Manufacturing's Sweeping Computer Revolution," *Production*, December 1979, p. 77.

67 Alan A. Hills, "Automated Maintenance Systems," in Machine Tool Task Force, *Machine Tool Systems Management and Utilization*, Technology of Machine Tools, vol. 2, p. 8.12-1.

68 Lesner, "Computerized Machine Monitoring," pp. 1–13.

69 Hills, "Automated Maintenance Systems," p. 8.12-3.

70 K. F. Noblitt, "A Computerized Approach to Planned Maintenance" (SME Technical Paper, MS78–481, 1978).

6: COMPUTERS OFF THE SHOP FLOOR: THE WIDER CONTEXT

1 Paul Kinnucan, "Computer-Aided Manufacturing Aims for Integration," *High Technology*, May–June 1982, p. 49.
2 Eric G. Petersen, "Increasing Productivity in Tool Design with Computer Graphics" (Paper presented at Worldwide Manufacturing Productivity Conference, General Motors Corporation, Warren, Mich., November 17–20, 1980), pp. 9–20.
3 "Tooling Built with Computer Graphics," *American Machinist*, November 1979, p. 92.
4 See Alice M. Greene, "CAD/CAM Shapes Up as Billion Dollar Market," *Iron Age*, February 23, 1981, p. 63; and Alice M. Greene, "CAD/CAM Systems Explode in Metalworking," *Iron Age*, December 28, 1981, p. 48.
5 Only those interviews not conducted by the author will be cited.
6 Report of a Working Party, *New Technology: Society, Employment and Skill* (London: Council for Science and Society, 1981), p. 41.
7 Ibid.
8 Howard H. Rosenbrock, "The Future of Control," *Automation*, vol. 13, 1977.
9 Mike Cooley, "Impact of CAD on the Designer and the Design Function," *Computer-Aided Design*, October 1977.
10 Mike Cooley, *Architect or Bee? The Human/Technology Relationship* (Slough, England: Langley Technical Services, 1980), p. 16.
11 Frederick J. Norton, "Interactive Graphics and Personnel Selection" (Paper in *Autofact West*, CAD/CAM VIII, Society of Manufacturing Engineers, November 17–20, 1980, Anaheim, Calif.), p. 114.
12 Quoted in Gene Bylinsky, "A New Industrial Revolution Is on the Way," *Fortune*, October 5, 1981, p. 110.
13 Joseph Harrington, Jr., *Computer Integrated Manufacturing* (New York: Industrial Press, 1973), p. 33.
14 Ibid.
15 Ibid., p. 3.

16 Ibid., p. 35.

17 Wickham Skinner, "The Stubborn Infrastructure of the Factory," presentation before Numerical Control Society, Seventh Annual Meeting and Technical Conference, Boston, April 1970, cited in Harrington, *Computer Integrated Manufacturing*, p. 34.

18 M. Lynne Markus and Jeffrey Pfeffer, "Power and the Design and Implementation of Accounting and Control Systems" (Monograph, Center for Information Systems Research, Massachusetts Institute of Technology, CISR No. 78, September 1981), p. 1.

19 M. Lynne Markus, "Implementation Politics: Top Management Support and User Involvement," *Systems, Objectives, Solutions* 1, 1981, p. 209.

20 Markus and Pfeffer, "Power and the Design," p. 8.

21 Markus, "Implementation Politics," p. 210.

22 Ibid., p. 210.

23 Ibid., p. 214.

24 Philip Caldwell, "A Strategy for the Eighties" (Speech to U.S. Steel Good Fellowship Club, Pittsburgh, January 20, 1983), p. 3.

25 Donald E. Petersen, "The Magic Machine at the Crossroads" (Speech at The Automobile and American Culture, Detroit Historical Society, October 1, 1982), p. 5.

26 Kenneth Gooding, "Ford Stakes Its Future on Erika," *Financial Times*, September 3, 1980.

27 Jerome Green, "Fisher Graphics Communications in Engineering" (Paper presented at Worldwide Manufacturing Productivity Conference, General Motors, Warren, Mich.), p. 9–2.

28 Ibid., p. 9–10.

29 Petersen, "The Magic Machine," p. 6.

30 Quoted in Bryan H. Berry, "Is Detroit Automaking Shrinking or Just Changing Its Shape?" *Iron Age*, July 27, 1981, p. 34.

31 Ibid.

32 Cited in "U.S. Automakers Reshape for World Competition," *Business Week*, June 21, 1982, p. 92.

33 Cited in "Latin America Comes of Age," *Automotive News*, February 9, 1981, p. 172.

34 U.S. Department of Commerce, "United States Automobile Industry: Status Report," submitted to the U.S. Senate Committee on Finance, Subcommittee on International Trade, December 1, 1981, p. 10.

35 This figure provided by UAW Research Department.

36 Marjorie Sorge, "Fraser Maps '82 Auto-Talk Issues," *Automotive News*, September 21, 1981, p. 1.

37 "Talk Now or Lose Jobs, GM's Smith Tells UAW," *Automotive News*, October 19, 1981, p. 3.

38 Donald E. Petersen, "The Future Task of the Worldwide Auto Industry" (Keynote address to *Automotive News* World Congress, Dearborn, Mich., July 19, 1981), p. 9.

39 Quoted in Raymond J. Larsen, "The World Machine: Can Manufacturers Make the Dream Come True?" *Iron Age*, August 18, 1980, p. 73.

40 Ibid., p. 69.

41 See Rosanne Brooks, "NMTBA Chief Foresees Offshore Units," *American Metal Market*, May 24, 1982, p. 5.

7: COMPUTERS AS STRIKEBREAKERS

1 Charles R. Perry, Andrew M. Kramer, and Thomas J. Scheider, *Operating During Strikes* (Monograph, The Wharton School, Labor Relations and Public Policy, University of Pennsylvania, no. 23, 1982), p. iii.

2 See Robert Lindsey, "Lines and Pilots Find New Setup Working Safely," *The New York Times*, August 13, 1981, p. 1.

3 Quoted in Edward Meadows, "The FAA Keeps Them Flying," *Fortune*, December 28, 1981, p. 51.

4 Robert Wesson, et al., "Scenarios for Evolution of Air Traffic Control" (Monograph, Rand, R-2698-FAA, November 1981), p. 23.

5 Quoted in Richard W. Hurd, "How PATCO Was Led into a Trap," *Nation*, December 26, 1981, p. 697.

6 Ibid.

7 Ibid.

8 Richard Witkin, "Futuristic Air Traffic Control System Pressed as Answer to Needs of Nation," *The New York Times*, April 18, 1982, p. 50.

9 Hoo-min D. Toong and Amar Gupta, "Automating Air-Traffic Control," *Technology Review*, April 1982, p. 54.

10 Richard Witkin, "Revamping of Air Control System in Next 20 Years Proposed by U.S.," *The New York Times*, January 29, 1982, p. 1.
11 "Automation Program Subjected to Test," *Aviation Week and Space Technology*, August 24, 1981, p. 25.
12 William H. Gregory, "Clearing the Air," *Aviation Week and Space Technology*, August 17, 1981, p. 13.
13 Toong and Gupta, "Automating Air-Traffic Control," p. 2.
14 Wesson, et al., "Scenarios for Evolution of Air Traffic Control," p. 2.
15 Ibid., p. 56.
16 Ibid., p. 57.
17 "Unions Hold the Computer Hostage," *World Business Weekly*, April 6, 1981, p. 8.
18 Barry Newman, "Computer Strike Snags a Bureaucracy," *Wall Street Journal*, May 19, 1981.
19 See A. H. Raskin, "The Big Squeeze on Labor Unions," *Atlantic*, October 1978, p. 41.
20 A. H. Raskin, "The Negotiation: Changes in the Balance of Power," *The New Yorker*, January 22, 1979.
21 This account is based on a series of interviews with Barry Fitzpatrick, a union officer at *The Times* (London) and a key coordinator of the strike.
22 This material on the Cadillac Seville is based on interviews conducted by the author, and is confirmed by "Cadillac Seville—How Small Was Made Beautiful," *Production Engineering*, January 1978, p. 70.
23 Perry, et al., "Operating During Strikes," p. 43.
24 Ibid., p. 111.

8: "A TECHNOLOGY BILL OF RIGHTS"

1 Alan Purchase and Carol F. Glover, "Office of the Future" (Stanford Research Institute, Business Intelligence Program, April 1976, no. 10), p. 5.
2 Undated Wang advertisement.
3 David A. Buchanan and David Boddy, "Advanced Technology and the Quality of Working Life: The Effects of Word Processing

on Video Typists," *Journal of Occupational Psychology*, no. 1, 1982, p. 1.

4 Barbara Cohen, "An Overview of NIOSH Research on Clerical Workers" in Daniel Marschall and Judith Gregory, *Office Automation: Jekyll or Hyde?* (Cleveland: Working Women Education Fund, 1983), p. 159.

5 Ibid.

6 Langdon Winner, "Do Artifacts Have Politics?" *Daedalus*, Winter 1980, p. 128.

7 Doris B. McLaughlin, "The Impact of Labor Unions on the Rate and Direction of Technological Innovation" (Institute of Labor and Industrial Relations, University of Michigan–Wayne State University, Report prepared for National Science Foundation, Grant PRA 77–15268, February 1979), p. vii.

8 I drafted this version of "The Technology Bill of Rights," with considerable input from Seymour Melman and the other attendees of the "Scientists and Engineers Conference," International Association of Machinists, New York, 1981.

9 See Hilary Wainwright and Dave Elliott, *The Lucas Plan: A New Trade Unionism in the Making?* (New York: Allison and Busby, 1982).

INDEX

Abernathy, William: *Productivity Dilemma, The*, 27
Adams, Al, 152–53
Adaptive control, 87–90, 91
Aerospace industry, 75, 92–93, 225, 259
Allis Chalmers, 149–50, 151, 152–53
AMC, 240
American Machinist magazine, 31, 75, 98, 131, 138, 155–56, 173
American Motors, 244
American Society of Mechanical Engineers, 21–22
APT (Automatic Programmed Tool), 99–100, 101, 122
Arnold, Horace A., 26
Assembly line, 26–27, 64
AT&T, 263
Attitude (worker), 92–93, 94
Authority, distribution of, 15, 190, 201–2. *See also* Managerial authority (control)
Auto industry, 3, 14–15, 26, 175, 185, 225; globalization of, 235–46; robots in, 168, 169
Automated industries: during strikes, 247–48, 256–59, 261–62, 263
Automated Manufacturing and Planning Engineering system (AMPLE), 176–77
Automation, 2, 4, 5, 6–10, 13, 45, 69, 137, 269; alternative view of, 274–78; democratic control of, 277–78; dynamic nature of, 121–22; limits to, 80–91; of office work, 266–67; of programming, 121–26; self-fulfilling prophecy in, 53–54. *See also* Computerization
Automation Systems, 10
Automatix, 161–62, 165
Autonomy, 4, 29, 31, 150–52, 190
Autoplace (co.), 172
Autoworkers, 162–63
Aviation Week and Space Technology, 252–53

Babbitt, Glenn D., 245
Barash, Moshe, 146, 150
Batch production, 137, 138, 181, 265
Beginner's Course in Numerical Control, 75–76
Bell System, 263
Bendix (co.), 164
Bennett, Harry, 189
Berdine, Robert A., 148
Beyer, Clarence, 86–87, 110
Black and Decker, 182
"Black book," 54–55, 139
Blue Shield of California, 258–59
Bluestone, Irving, 105–6
Blumberg, Melvin, 147, 148–49, 151, 154
Boeing, 221

Borzcik, Paul S., 145–46, 147–48
Breakdown(s), 5, 147, 185
British Leyland, 244
British Overseas Corporation, 112
Busch, Thomas J., 241–42
Business Week, 169

CAD-D, 122
Caldwell, Phillip, 235
Calma, 10, 165
Carter, Charles F., Jr., 178
Caterpillar Tractor, 75, 83, 93, 95, 96–97, 111, 149
Center for Policy Alternatives, MIT, 74
Cessna Aircraft, 112
Charles Stark Draper Laboratories, 142
Chern, Bernard, 60
Chrysler, 160, 166–67, 240–41, 243
Cincinnati Milacron, 75, 164
Civil service unions (Great Britain), 255
Collective action, 28, 33–40, 41
Collective bargaining, 33, 95, 270–71, 273–74, 276
Communication Workers of America (CWA), 263
COMPACT II (programming language), 100, 102, 103
Complexity, 5, 57–58, 125, 147, 148
Computer-aided design (CAD), 122, 168, 218–27
Computer-aided design and computer aided manufacturing (CAD/CAM), 2, 165
Computer-aided process planning (CAPP), 55
Computer graphics, 221
Computer-integrated manufacturing (CIM), 2
Computer Integrated Manufacturing, 139–40
Computer networking systems, 9–10, 137, 138–40, 160, 175–90, 266, 268; and power relations, 191; robot in, 161–62, 172
Computer numerical control (CNC), 70, 98–126
Computerization, 1, 22, 55, 73, 74, 126, 217; control factors of, 48, 229–31; global factory, 234–46; in metalworking, 70–75; of process planning, 55; positive potential of, 267–68; and shortage of skilled labor, 53–54; social purposes of, 10–11; utilizing human potential, 64–65. *See also* Automation
Computers, 6–10, 12, 15, 61, 68, 225; information-gathering potential of, 2, 8–9; and managerial structure, 227–34; in organization of work, 4, 26, 264, 267–68; as strikebreakers, 247–63; surveillance possibilities of, 109; in wider context, 217–46
Condec, 165
Consight, 159
Control, 9, 137; through computer technology, 229–31; of manufacturing process, 44, 45–65; through NC, 66–67; through robots, 171–72; of technology, 251, 255. *See also* Adaptive control; Managerial authority (control); Worker control
Cooley, Mike, 222, 225–27
Council for Science and Society (England), 59, 62, 63, 223–24

Craft: standardization of, 197–98. *See also* Pride of craft
Creativity, 226, 268, 269
Cross Company, 245

Daley, Frank, 230–31
Daley, Jim, 132
Dallas *Morning News*, *Times Herald* strikes, 257
Deaton, Homer, 260
Decision-making, 15, 60–61, 177, 226–27; democratic, 4, 27, 46; locus of, 5, 120, 124; production, 61, 63–64, 70, 98
Defense Science Board, 52
Dennen, Irwin H., 245–46
Design, 4, 5, 15, 72, 85, 98, 219; of computer systems, 153–54; computers and, 12, 217, 218–27; criteria of, 45–65, 67–68, 69; global factory, 238–39; for human input, 126, 142, 146–47; of information systems, 177–78. *See also* Computer-aided design (CAD)
Diemakers, 29–44
Diemaking, 90–91, 190, 191–216
Digital Equipment Corporation, 164
Direct numerical control (DNC), 138–40, 181–82
Discipline, 20, 38, 43, 150, 177; through globalization, 243; information management systems and, 181, 183; robot use in, 171–72
Displacement, 135: by robots, 156–57, 168–70
Division of labor, global, 245–46
Donovan, Jack, 35, 36–37, 196–97, 199, 204, 208, 211, 212–13
Downtime, 85, 178, 180, 188, 189; FMS, 146, 148–49, 150
Dual-sourcing, 240–41
Dubraski, Ben, 129
Dunlop, John, 251

Economies of scale, 163, 240
Employment, 3, 134–35, 155, 157, 168–70
Engelberger, Joseph, 156, 162, 167, 171
Engineering ideology, 45, 46, 58–59
Engineers, 101, 222
Experience, 60, 63, 81, 83–84, 166–67
Export platforms, 240, 242

Factory, computerized, 2, 12, 56, 66, 136–216. *See also* Global factory
Fairchild Aircraft, 111
Fairn, Ken, 82
Fast Cut (system), 122–23
Faurote, Fay L., 26
Federal Aviation Administration (FAA), 249, 252, 253, 254
Federal funding of technology, 276–77
Feedback, 9, 68, 85–87, 139–40, 177
Fenton, George, 183
Fiat, 160
Financial Times, 238
Flexibility, 1, 6–7, 32, 140–41, 158; of NC, 72, 77, 120
Flexible manufacturing systems (FMS), 137, 138–55
Ford, Henry, 26, 29, 175
Ford Motor Company, 33, 64, 83, 178, 189, 237; Batavia plant, 174–76, 178–79; Rouge complex, 29, 31–32, 41, 44, 64,

Ford Motor Company (*cont'd*) 109, 190–216; world-wide car factory, 236, 238, 239–40, 243, 244–45
Foreman, 31, 36–37, 182–83
Frost and Sullivan, 242

G and T Products, 112–13
Gallmeyer, William C., 246
Gardner, Al, 31, 35, 36, 42, 43–44, 208, 212, 213, 214
General Electric, 2, 3, 10, 111, 181–82, 259–60; robotics, 164, 165, 169
General Motors, 90, 97, 164, 185, 260–61; international operations, 239–40, 243; PUMA system, 173–74; robotics, 158, 159, 160, 162–63, 164, 165–66, 168–69
General Numeric, 108–9
Gerwin, Donald, 144, 147, 148–49, 151, 154
Global factory, 234–46
Globalization, 242–43
Gossard, David, 120–21, 223, 224, 226
Green, Jerome, 225, 238–39
Greening, John H., 73
Grievances, 32, 40–41, 79, 104–5, 115
Gupta, Amar, 252, 253
Gustafson, Burnell A., 92

Haas, Paul R., 145
Halevi, Gideon; *Role of Computers in Manufacturing*, 60
Harrington, Joseph, 67, 228–30
Health and safety concerns, 79–80, 132, 152, 267. *See also* Stress
Hedden, Russel A., 49–50
Heide, Stan, 195–96, 198, 202–4, 205, 206, 207, 213–15
Helms, J. Lynn, 251, 252
Hills, Alan, 187, 188
Honda, 156
Hopwood, Norm, 125, 179, 180, 184
Howard, Dennis, 101–2
Human input, 5, 58–60, 91, 126, 137; air controllers, 249–50, 253–54; elimination of, 11, 46–47, 60–62, 125, 265; need for, 14, 62, 68–69, 92. *See also* Skill(s), elimination of; worker input

IBM, 164, 177, 220
Independence (on the job), 28, 31–32, 42–43, 185–87
Industrial Machine and Engineering, Inc., 113
Information, 8–9, 139, 202, 227–28, 229, 266; control of, in TOPS, 211–12, 213–14; is power, 228, 232–33. *See also* Management information systems (MIS)
Information processing, 66–67, 137
Integrated Computer-Aided Manufacturing Program (ICAM), 55–57
International Association of Machinists (IAM), 78, 113–18, 270
International Association of Machinists and Aerospace Workers, 271
Intersil (co.), 10, 165
Iron Age magazine, 47, 49, 87, 145, 154

Jansen, Ralph, 182, 183–84
Japan, 156–57, 237, 239

Japan Industrial Robot Association, 156
Jessen, Jack, 123–25, 229
Job creation, 54, 57, 265
Job loss, 32, 54, 79, 135, 137, 142, 157, 167–70, 196–97. *See also* Employment
Job satisfaction, 150–51
Job security, 32–33, 42, 270
Johnstone, Richard, 88
Jones, Peter, 255

Kansas City Star strike, 257
Kawasaki, 156
Kearney and Trecker, 109, 145
Keno, Doug, 51, 83, 85
King, Bob, 188–89
King, R. E., 103
Knowledge, 59–60, 62–63: definition and monopolization of, 22–23, 24, 54–55. *See also* Experience; Information
Kuhn, Ralph, 195, 204–5

Labor, 27, 234; direct, 142, 143, 185; discipline of, by globalization, 243. *See also* Discipline; Unions
Labor-management relations, 12, 58, 254
Labor solidarity, 258, 262
Laverty, Chris, 129–31
Leaders, 34–37; and TOPS, 196–204, 211–16
Leone, Russ, 30, 31
Lesner, William T., 182–83
Leverage (of workers), 14, 43, 80, 101, 269; through collective action, 33; information as, 202; skill as basis of, 68, 259; limitations on, 49; of unions, 247, 261–63
Lewis, John L., 262
Liscinsky, Stephen, 102
Local-content laws, 239, 240
Lockheed-Georgia, 53, 139
London Times strike, 258, 262
Lucas Aerospace, 274–76
Lucas Industries, 274–76

McCoy, Gary J., 178
McDonnell Douglas, 78, 122, 123, 129, 181, 259; F-15, 72; plant, St. Louis, 8, 17
MacMillan, S. M., 178
Machine Design, 158–59
Machine monitoring system, 177–90
Machine shop, 16–44
Machine Tool Task Force, 46, 52, 60, 122
Machinists, 17–20, 24, 28–29, 126; impact of NC on, 70, 126–35; as programmers, 101–3, 106–7
Maintenance monitoring systems, 177–90
Malingering. *See* Soldiering
Management, 25–26, 44, 45–46, 50–53, 167; control of technology, 270, 274; and NC, 70, 73, 76, 82; resistance to TOPS, 205–9; rights of, 4, 27, 80, 104–5, 269, 274. *See also* Scientific management
Management information systems (MIS), 12, 137, 138, 174–90
Management structure, 12; computers and, 217, 218, 227–34
Managerial authority (control), 4, 11, 39, 48–49, 67, 218, 222, 264, 268; and automation of shop floor, 138, 153–54, 177–78; CAD and, 221–22; effect on technology, 13, 14–15;

Managerial authority (*cont'd*)
fragmentation of, 228–29; redistribution of, 205–6; technology in conflict with worker control and, 70, 82, 84–85, 86–87, 90, 92, 96–97, 98, 102–3, 107, 111–12, 113–20, 265–66
Manual data input (MDI), 108, 109
Manufacturing, 2–3, 12, 61, 227–28
Manufacturing Data Services Incorporated, 100, 102, 103
Manufacturing resource planning (MRP), 177
Markets, international, 235, 236–37
Markus, M. Lynn, 231, 232, 233–34
Mass production, 22, 71, 138
Massachusetts Institute of Technology, 135, 170
Massey Ferguson, 83
Mathues, Thomas O., 157
Matthews, Dave, 43
Maxwell Motor Car Co., 160
Mayo, Elton, 27
Medium-batch production, 11–12, 83, 138
Messerschmitt-Bolkow-Blohm, 141, 146
Metalworking, 20–21, 23–24, 27–28, 62–63, 68, 126; computerization of, 70–75
Metalworking industry, 46, 154
Microelectronics, 1, 2, 4, 6, 12, 15, 68, 235; in organization of work, 264, 267–68
Military (the), 55–57, 73, 276
Military aircraft, 72
Mirkava tank, 65
Modern Machine Shop magazine, 86, 98
Monitoring, 9, 14, 25, 177–90; in TOPS, 196, 199–200, 201–2, 203
Montgomery, David, 28–29
Moog Hydra Point, 53
Moore Special Tool Company, 102
Moral code, 28–44
Moriarity, Brian, 142–43
Motivation of workers, 27, 49–50, 92–93, 96
Multinational corporations, 3, 9, 166, 234–35

National Institute of Occupational Safety and Health (NIOSH), 267
National Machine Tool Builders show, 107
National Science Foundation, 270
National Tooling and Machinery Association, 52
New York newspapers strike, 257–58
Newspaper industry, 262
Nissan, 240, 244
Noblitt, K. F., 189
Nordhuff, Bob, 103, 125
Norton, Frederick J., 227
Norwegian Employers' Federation, 276
Norwegian Federation of Trade Unions, 276
Nulty, Leslie, 84–85, 87, 113–14
Numerical control (NC), 11–12, 47–51, 66–135, 136, 137, 138, 139–41, 142, 145, 157, 168, 173, 181; diffusion of, 55–56; problems with, 73–74, 106–7; technical and economic advantages of, 70–75; undermines leverage, 259–61. *See also* Flexible manufacturing systems (FMS)

Numerical Control Society, 48, 111
Numerically controlled machine tool(s), 1, 265

O'Brien, John, 192–93, 194, 195, 196, 197, 201–2, 203, 206–8, 209, 215
Office and Professional Employees Union (OPEU), 258–59
Office of Technology Assessment, 169
Office work, 265–67
Ohmae, Kenichi, 171–72
Operating During Strikes (monograph), 248, 261–62
Outsourcing, 235, 243
Override switch (NC), 86–87, 88, 91, 139

Pace of work, 14, 42, 78–80, 88–90, 172–74, 180
Palmer, Richard, 82
Part programming, 47, 67, 73, 81–82, 85, 98–99, 120–21, 259
Pestillo, Peter, 243
Petersen, Donald E., 236–37, 240, 244
Peti, Steve, 113
Peugeot, 244
Planning/operation separation, 45–46, 67–68, 85, 125–26, 147–48, 152, 222–24
Planning department, 24–25, 55, 63
Polcyn, Stanley, 168
Power, 4–5, 12, 15; and computerization, 82–85, 126, 230–31; information as, 228, 232–33; issue with skilled workers, 18, 20–28, 188–89
Power relations, 32–33, 191, 268; in management, 231–34; with TOPS, 215, 216
Price, David A., 81
Pride of craft, 28, 30–31, 33, 41, 130, 186
Process planning, 54–55
Production, 4, 5, 7, 8–9, 11, 35, 175, 234, 264; alternative mode of (proposed), 275–76; computer analysis of, 180–82; global, 12, 217, 234–36
Production magazine, 172
Production managers, 190, 216
Productivity, 5–6, 27, 39, 46, 63, 82, 142, 218; effect of technology on, 146, 268, 269; with information management systems, 189–90; through numerical control, 68, 71, 75; robots and, 155–56
Professional Air Traffic Controllers Organization (PATCO), 248–54, 258
Profitability, 4, 5, 46, 57, 155
Program editing, 110–11, 112, 119–20
Programmable controllers (PC), 175–76
Programming, 70, 98–126; location of, 107–12
Programming languages, 67, 99–101
Progress, ideology of, 127–28
PROMPT (programming language), 100–1
PUMA (Programmable Universal Machine for Assembly), 173–74
Purdue Univ., School of Industrial Engineering, 136–37

Quality of work life, 27, 63, 170–71

Raibert, Marc, 159–60
Rand Corporation, 249, 253–54
Raskin, A. H., 256–57
Rationalization of workplace, 26, 74. *See also* Scientific management
Reagan Administration, 248, 251
Reardon, Dennis, 251
Reicher, Mike, 133
Reliability, 13, 14, 57, 155; of robots, 163–64
Renault, 164, 244
Repair of machinery, 185–89
Resistance, 123–24, 127; to computerization, 231–34; to management information systems, 184, 188–89; to robots, 167; to TOPS, 204–9, 210–16
Robogate, 160–61, 166–67
Robot VI exhibition, 165
Robotics, 156, 164–74
Robots, 2, 7, 12, 137, 138, 155–74, 270; cost of, 162–63; constraints on use of, 166–67; defined, 157–58; first-generation, 158–59, 164, 166, 170–71; second-generation, 171; visual/tactile sensing, 159–60, 169
Rockefeller, David, 57
Rosenbrock, Howard, 224–25

Saab Scania, 58
Schenck, Douglas, 88
Schloss, Edward F., 102
Schlumberger (conglomerate), 100
Schwartz, Robert, 76–77
Schwartz Machine Company, 76–77
Scientific management, 11, 20–28, 101. *See also* Taylorism
Seniority, 34, 51, 95–97
SETUP (Shop Entered Tutorial Programming), 109
Shifo, Thomas P., 143
Shop floor; computerization of, 136–216; NC on, 70, 75–97
Skill(s), 4, 5, 12, 17–18, 265; codified, 67–68; elimination of, 26, 63, 68–69, 70; feedback, 85–87; knowledge underlying, 59; leverage through, 259; minimization of need for, 46
Skill availability, 68, 94
Skill creation, 57
Skill erosion, 123–24, 190, 222
Skill level, 14, 113; needed with NC, 92–94, 133; reduction of, 33, 46–48, 52, 57, 67, 75–78, 80–91, 123–25, 187; and TOPS, 190, 192–93
Skilled workers, 52–53, 101, 185–90
Skinner, Wickham, 230
Small-batch production, 11–12, 44, 138
Smith, Donald, 102–3
Smith, Oberlin, 21
Smith, Roger, 169, 243, 270
Social goals of automation, 45–47, 66, 67–69, 98, 190; shop floor, 138, 142, 145–46
Social impact of robots, 168–71; of scientific management, 26–27
Society: costs to, in automation, 5, 13
Soldiering, 23, 89
Stanford Research Institute, 266
Stanovsky, Clint, 83, 93, 95, 96
Stress, 5, 85, 130–31, 151, 222, 226–27, 267
Strikebreakers: computers as, 247–63

Strikes, 246
Superautomation, 6–10, 153–54, 266
Supervision, 21, 42, 43, 183–84, 185–87
Supervisors, 132, 167; in strikes 250, 256–57, 259–60
Sutton, George P., 52
Swedish Employer's Confederation, 58

Tapocik, George, 34–35
Taylor, Frederick W., 21–28, 44, 48–49, 55, 61, 62–63, 89, 193, 265
Taylorism, 38, 55, 185, 225, 228
Technical goals of automation, 45, 142–45
Technological change, 1–15, 155, 157, 264–65; inevitability of, 127–28, 269–70
Technology, 44, 64, 95–96, 164–66; in conflict with worker control and managerial authority, 70, 82, 84–85, 86–87, 90, 92, 96–97, 98, 102–3, 107, 111–12, 113–20, 265–66; democratization of, 13, 15; embodying human values, 274–78; federal funding of, 276–77; for its own sake, 73–74, 141–42; goals in shaping of, 251–52, 264, 268–69; limiting possibilities of, 219, 221–22, 224–25, 226–27; in organization of work, 47–65
"Technology Bill of Rights," 264–78; text of, 271–73
Technology Review, 252
Telecommunications, 9, 12, 220, 234, 238, 247–48
Telephone Workers Union (TWU), 262–63
Texas Instruments, 164
Time-and-motion study, 23, 25, 26
Time study, 179, 186, 189; TOPS as, 196, 198–200, 202–3
Toong, Hoomin D., 252, 253
Torendeau, Jean Claude, 143
Total operations planning system (TOPS), 190–216, 230
Towne, Henry R., 21
Toyota, 156
Trapoti, Greg, 130
Tulkoff, Joseph, 139

Unemployment, 269
Unimation, 156, 164, 165, 172
Union(s), 4, 13, 14, 37, 40–41, 69, 94–95, 124; leverage of, in strikes, 247, 261–63; stake in technological change, 103–7, 269–70, 274, 276–78; and TOPS, 214, 216. *See also* Grievances
U.S. Air Force, 46, 55–57, 73, 100
United Auto Workers, 4, 29, 33, 104, 105, 243, 270
U.S. Department of Defense, Manufacturing Technology (Man Tech) program, 276
U.S. Department of Transportation, Transportation Systems Center, 242
U.S. National Commission on Technology, Automation, and Economic Progress, 66
United Technologies (co.), 164
Univ. of Michigan, 242

Van Vuren, Raymond, 249
Villers, Philippe, 161–62
Vining, Richard, 161
Volkswagen (co.), 164, 240
Volvo, 244; Kalmar plant, 58

Wage rates, 170, 234, 240
Wang (co.), 266
Warner Gear, 183
Washington Post strike, 256–57
Westinghouse, 164, 165
White-Sundstrand Machine Tool Company, 138–39
Williams, Cas, 259
Winner, Langdon: "Do Artifacts Have Politics?" 268
Wisnosky, Dennis, 228
Word processors, 168, 266
Work, 15, 23; degradation of, 5, 61, 63, 68; subdivision of, 14–15, 22, 25, 26, 228
Work environment, 15, 46, 142
Work ethic, 49, 51
Work organization, 2, 4, 70, 126, 153, 157, 264, 265–68; alternative possibilities in, 58, 111–12; CAD and, 219–20; power in, 32–33; technology in, 13–14, 46–65
Work pace. *See* Pace of work
Work relations, 42, 183–84
Work rules, 32–33, 34, 43, 235
Work to rule, 19–20, 186–87
Worker control, 27, 28, 44, 70, 80–91, 98, 107–8; knowledge in, 54–55
Worker input, 5, 19, 57, 58, 67, 264; CNC, 109–10, 111–12; elimination of, 142, 147; need for, 38–40. *See also* Human input
Workers, 5, 6, 9, 27, 58, 64, 178, 222; fears of new machines, 134–35; impact of robot use on, 167–74; tactics to oppose management, 33, 38, 179–80, 184–85. *See also* Machinists; Skilled workers
Workplace, 3–4, 13, 15, 45, 48–49, 145, 265; organization of, 26, 74, 98, 107; two-tiered, 57